HOUSE, PLUS ENVIRONMENT

RONALD L. MOLEN

Olympus Publishing Company ⊛ Salt Lake City, Utah

ISBN 0-913420-17-4

Library of Congress Catalog Card Number: 73-83499

CONTENTS

Foreword 5

Acknowledgments 7

Man's Need for Housing 9

Housing in America 25

Application of the Theory 39

Residential Architecture 59

Decorating and Landscaping 73

Appliances, Materials, and Technology 91

The Housing Industry 109

The Neighborhood 129

Educational vs Pseudo Communities 153

A Day in the Village 179

Conclusion 193

Appendix 197

References 203

Index 207

Foreword

F rom simple beginnings — a cave, a hut, an igloo, an animal-hide tepee — the "house" has always been the emblem of man's place in the scheme of things. Contemporary man no longer looks upon his house as just an abode...rather, he views it as "home," his status symbol, interpreting to himself that it and all that accompany it on his property depict to the world his position in life, indeed even his wealth.

Whether that house is a stereotype of many others on his street or in his neighborhood, whether it is small or large, whether it sits side by side with others of its ilk on narrow lots or is placed in a grander setting of manicured lawns and gardens, or whether it is in a forest or on rangeland, it is still his personal possession. It is *home*.

Mr. Molen's deep concern about the effect of the house itself upon the quality of the interactions within the family, and in turn the interactions of the family within the community, has prompted the writing of this book. As an architect, he would propose a concept of total living — not just housing per se.

In *House, plus Environment*, Mr. Molen points out that, considering the obvious influence of physical surroundings on the types of individuals our society will produce, the housing industry has "clearly abdicated its responsibility" by offering "shoddy and shallow" solutions to the complex and vital problems of human growth and interaction. He believes that this has happened because there is a great void in the basic philosophy of what a house should provide. The void, he feels, is "partially and inadequately filled with foolish elements that do more to destroy family interaction than to enhance it." He also feels that the possibilities for creating a lush and exciting residential environment are not only worth striving for, but in a period of social turmoil and unrest, become imperative.

The book is addressed to two groups: the great numbers of people involved in deciding what kind of housing is produced, and those who are deeply concerned with providing the best possible environment for their families.

House, plus Environment places the reader foursquare on the threshold of a broader concept in home ownership — that of owning not only a house, but also the significant environment surrounding that home.

The Editors

Acknowledgments

First, I must acknowledge the patience of a devoted wife and the endurance of three children for Sundays lost and vacations consumed in gathering source material. Also, I am indebted to the many people who helped unravel the confused ramblings of an often awkward but always inadequate writer.

I must express my deep appreciation for the opportunity of working with Research Homes and specifically with Vern Hardman for being the always hoped-for client who not only allowed experimentation, but encouraged it despite an occasional economic loss. Also, I would like to acknowledge the help of the many excellent craftsmen in the company who not only contributed greatly to the quality of the product, but also made the process of producing it most enjoyable.

Not to be forgotten are parents who provided a vital home environment for me and who are now involved in contributing much to the quality of life of their grandchildren.

Three other factors should also be mentioned. Living in a tight, well-functioning neighborhood in a suburb of Chicago (that was my good fortune to experience as I was growing up), living three years in Austria and Switzerland in well-formed communities in urban centers, and being exposed to a basic religious heritage collectively have influenced the concepts in this book.

Finally, I must acknowledge the excellent writers and scholars who not only made research for the book exciting and personally rewarding, but who also are responsible for many of the significant elements of the book.

CHAPTER ONE

Man's Need for Housing

"In numberless animal societies, the struggle between separate individuals for the means of existence disappears; struggle is replaced by cooperation."
Charles Darwin

Housing is a basic element in the structure of our society. The home has a great impact on the growing child — an impact unquestionably more important than that of either the school or church. The physical home can provide privacy and intimacy, can encourage exploration, self-determination, and creativity, and can help to develop emotional harmony and love of beauty. A well-designed house will never be a substitute for vitally concerned parents, but it can provide the right kind of space for the right things to happen.

The single-family dwelling is an ideal place in which to raise a family. But what is the real significance of the single-family house, and how should it relate to the neighborhood and the community? Let us explore some exciting new concepts in the fields of biology, sociology, anthropology, and psychology. Many of these ideas remain at the theory level, but there is nevertheless an immense amount of material to substantiate each position. We are seeking a scientific view of man, the ingredient of common humanity in the primitive *Homo sapiens*, the product of thousands of years of evolution. We are searching for something vital — the instinct, the structure, the idea, or the relationship through which man can integrate successfully with his environment.

Territory and Pair Bond

Sigmund Freud maintained that the sex drive is the most significant biological demand and therefore the prime force influencing man both physically and psychologically. In contrast, the "new biology" maintains that the sex drive is subordinate to the biological need for territory and that this inexorable pattern rules an enormous variety of species, including man.

In most species of animal, the male establishes the territory. He either rounds up his females, or they arrive voluntarily for mating. Many animals pair off — one male to one female. The primary function of the female is to provide the security and education of the young, while the male guards the territory required for them to sustain life. The male continues to function as guardian until the survival of the young is virtually assured. Since this length of time is reasonably short, animals do not sustain a single relationship throughout their life.

For man, the maturation period takes eighteen to twenty years. In that length of time, a strong "pair bond" between the male and female develops. This pattern is a necessary protection for the young, for only in a well-protected territory ruled by a congenially bonded pair will the offspring

Man has a biological need for territory.

A strong "pair bond" between male and female develops.

In a well-protected territory the offspring find security.

find the security, the stimulation, and the personal identity so crucially needed during the period of maturation. Our society is changing at an alarming speed, yet the slow evolutionary process responsible for man's genetic inheritance maintains man as a constant, and whatever radical changes occur should be designed around him. He simply does not change rapidly.

The modern, more permissive mores may thus violate a significant biological pattern deeply imbedded in the human psyche throughout those thousands of years of evolution. A too-rapid and perhaps unnatural change is illustrated in the following example known to have happened to an acquaintance who, with his wife, joined a "commune" in the hope of finding a new and "more valid" way of life. In the commune, sexual freedom among all members was considered a modern innovation to be practiced openly. Eventually all sexual relations ceased, even between married couples, and in many cases marriages were terminated. The intellectual analysis of the group was that cultural "hangups" were responsible; but it may be that the members violated a basic biological truth: that the function of the pair bond is to provide a secure environment for the young, and that a private territory is required to sustain the physical and psychological needs of the young. Promiscuity is not just a violation of God's law in the Judaeo-Christian tradition, but is a violation of natural law and its mechanisms to ensure the species' survival. Territory therefore achieves immense significance as the minimum requirement for the proper functioning of the pair bond.

11

Enmity is a direct consequence of territoriality, for one's neighbor is his natural enemy.

Enmity is replaced by cooperation when outside hazards develop.

If territory is indeed the fundamental animal need, what kind of territory is correct for man? What are the necessary conditions of territory for the optimum functioning of the human pair bond? Can the territorial instinct survive the mobile home on its *rented* space? Does a subsidized apartment in a high-rise building fulfill that need? Can the male in the pair bond satisfy his territorial need in a rented, high-rise apartment? The answers to these and other questions are immensely important; but only time will provide complete answers. Yet we *do* have enough information to establish some very strong directions. We also have enough information to know that we are actively engaged in producing housing that in many ways violates man's basic biological needs.

Another dimension of territoriality is the nonexclusive or cooperative territory (often referred to as "social space") where animals work together in groups. This group territory seems to be held together by the formula developed by Robert Ardrey in *The Territorial Imperative*:

$$A \text{ (amity)} = E \text{ (enmity)} + H \text{ (hazard)}$$

Amity exists in a group when enmity and hazard achieve the proper balance. Enmity is a direct consequence of territoriality, for one's immediate neighbor is his natural enemy since both are recipients of the frictions generated by a common boundary. Enmity is only reduced when some hazard

12

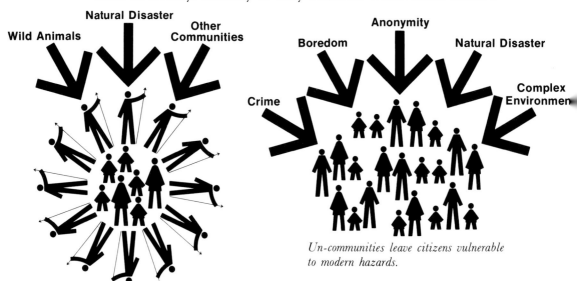

Hazard was the binding force of primitive communities.

Un-communities leave citizens vulnerable to modern hazards.

becomes so overwhelming that the warring neighbors must stand together. The territory must be fought for collectively to ensure survival. As the hazard approaches 100 percent, and enmity is therefore reduced to zero, the ideal conditions of amity exist.

The cooperative or group territory has a most significant contribution to make. Here individuals learn the advantages of cooperating and working together, realizing that certain benefits can only be achieved through common effort. Group territory, then, becomes an extremely important element central to community, and the security, identity, and stimulation so crucial to a healthy environment can only be achieved where a healthy community exists.

We do not know how many of our modern "luxury" suburbs are what has been termed "*un*-communities"; nor do we know how many people live in total isolation. The numbers of children who grow up in families turned in on themselves, leading self-centered, narrow, and truly miserable existences, are unknown to us. Many crimes are committed by people totally ignorant of any responsibility to the community, and other crimes are perpetrated by gangs of youths wherein the hazard of survival is the binding force of a small group waging "war" on a larger, anonymous community. Yet in our nation today many small towns and communities still have a vitally functioning cooperative territory.

13

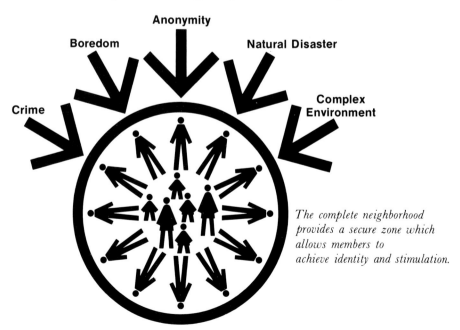

The complete neighborhood provides a secure zone which allows members to achieve identity and stimulation.

Thus it becomes apparent that *private* territory plus *cooperative* territory are imperative to a healthy society, and that it is better for children, and indeed all family members, to live in a humble dwelling in a vital community than in a mansion in an incomprehensible *un*-neighborhood.

A child must grow up in and be a participant in a healthy society if he is to become an adult who understands democratic processes, who feels a distinct personal commitment to the society at large. Suburbia, with its pitifully weak societal structure, has produced several generations of individualistic consumers, too often lacking in any commitment to the quality of society in which they participate. The tragedy is that too much of America lives not only in substandard housing, but in un-neighborhoods as well.

Orbit and Scale

Where we live, work, socialize, and play — in other words, the general configuration of our daily activities — comprises our *orbit*. We say that we are citizens of our city, but in a very literal sense, this is an impossibility. An entire city is so vast and incomprehensible that it is beyond our capacity to respond to it as a whole. We can only truly respond to stimuli in our orbit. We can only be true citizens of our neighborhood community. We can only relate to a limited number of people.

When animals are forced to relate to unnatural and excessive numbers of other animals, a "behavioral sink" develops. John B. Calhoun, in *The Role of Space in Animal Sociology*, tells us that:

> In a state of nature [overcrowding] may cause at least seven times the optimum number of animals to assemble at one place, with a resulting accompanying array of abnormal behavior developing. Prominent among these are nearly total dissolution of all maternal behavior, predominance of homosexuality and a marked social withdrawal to the point where many individuals become unaware of their associates despite their close proximity.

Huge cities with no neighborhood structure, where housing exists in incredible densities, will inevitably express the salient

14

characteristics of a behavioral sink. Again, we must accept man for what he is by not violating his natural biological needs. Dissolution of maternal instincts, homosexuality, and social withdrawal are simply nature's way of reducing the population to a more manageable size.

Hubert H. Humphrey described his vision of America as a nation of villages "not only in Montana, but Brooklyn, where people live and work and have their own police, hospital and schools, where a death is known and grieved from one town to the other, where a valedictorian is honored by all."

The function of planning, then, is cognizance of scale and the realization that man has characteristics and limitations that must not only influence but also determine design. Critics say that it is an oversimplification of planning to define a healthy city as simply a collection of healthy neighborhoods. Even so, the village or neighborhood ideal, based upon concepts of fit and orbit, does offer a sound basis from which to attack many complex problems. And the maintenance of proper scale can ensure a higher quality of life.

The function of the small community then takes on monumental significance. It is the critical survival mechanism through which the value system of society is perpetuated. School and church must reinforce this process, but we must recognize the limitations of these remote institutions as contributions only and not the community process itself.

15

Man is programmed for living in communities of limited size (scale) and limited numbers (density). Scale and density limit him to a specific orbit.

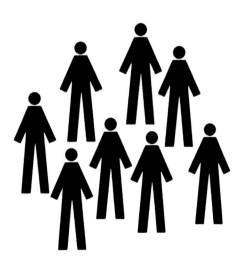

Society must continually renew itself, and the community is the level at which these slow evolutionary processes should occur if the society is to achieve fit.

"Fit" or Appropriateness

Anthropologists study the basic living patterns of an ancient people and then try to determine the appropriateness of the societal structure which evolved. This appropriateness is called "fit." Anthropologists have found that often extremely primitive societies developed excellent fit. As a society evolves, however, and power becomes centralized, the fit begins to deteriorate because even more arbitrary and remote decisions are made. Society itself can deteriorate as a result and lose its power as a civilizing force, becoming responsible instead for some grossly uncivilized actions (the classic example being, of course, the attempted genocide of the Jews by the Nazis). When the decision making remains local — a matter of tribal law — the societal fit remains good for long periods of time, and the society thus produced has a healthy foundation.

Many utopian schemes for the ideal "fit" have been proposed, but they have generally been arbitrary thoughts of a solitary genius who almost invariably has not been willing to accept man as he is. The utopian idealist has wanted to remake man in the image of his vision of Utopia, and has thus

16

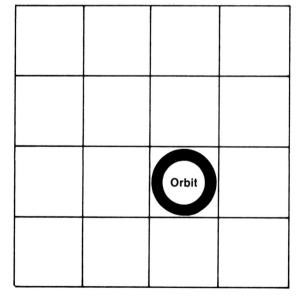

A neighborhood is a grouping of people who share an orbit. A person is first a citizen of his neighborhood, and through his orbit he relates to the city at large.

forced the fit. A society recognizing man for what he is might never achieve the highest utopian standards, but such a structure certainly would provide a realistic base from which to work step by step toward the great possibilities of the future.

Paul Leyhausen, an eminent biologist, has said:

> Modern psychology and sociology have for far too long been obsessed by the ideal that maladjustment between individual and society is due to a faulty construction of the individual who must, therefore, be helped to adjust to the demands of a society which is taken as a more or less unalterable system of conditions. In the present situation, this is decidedly the wrong way of looking at the problem; as history clearly teaches, societies and their structures have undergone rapid changes all the time, and there is no reason to assume that adaptive changes could not be effected by conscious human effort. For practical purposes, and in striking contrast to common belief (even of the scientific community), the limits within which the individual can adapt and stay healthy are rather narrow and cannot be changed without interference with the basic pattern of human nature itself, that is, without danger of destroying the species. We should therefore stop striving vainly to adapt the individual to the impossible demands of a society which regards itself as an end instead of a means to a better and happier life for the individual.

17

In the frenetically changing contemporary world, plagued by "future shock," only primary groups with the ability to rapidly change the societal structure to the needs of man will be able to provide a healthy basis upon which community life can continue. Society must continually renew itself. A vital, cooperative territory must have fit, and the community is the level at which these slow evolutionary processes should occur if the society is to achieve fit.

Behavioral Modification

The behaviorist school of psychology stresses the immensely significant conditioning that occurs in environment. What we are because of our heredity and our biological past is considered to be less significant. The behaviorists feel that proper conditioning can determine one's finally formed personality. This is accomplished through rewarding or positively reinforcing those qualities that "ought" to be integrated into the person. This of course sounds much like the "brave new world" and to many is a somewhat frightening prospect.

The extent to which behavioral conditioning can actually be applied and be effective is still a debatable point, but the important thing is that the kinds of stimuli that are constantly bombarding us, and to which we are constantly responding, do have great influence upon us.

Many years ago, when the advisability of city planning was still being debated, critics were asking what was really wrong with laissez-faire development — growth that occurred sporadically, directed by unconscious socioeconomic forces. We developed a kind of blind faith that what was good for business would somehow prove beneficial to all. This simple faith of course has been shattered. Today we know that planning can have minor negative aspects, but certainly few are still disputing the critical need for better planned cities.

We are still at the mercy of myriad stimuli, both positive and negative, produced by the same unconscious socioeconomic forces. We have never been enchanted with the idea of an exhaustively designed culture that would provide our decisions as to which are good and bad stimuli, because such an approach would unquestionably lead to a highly censored society directed by a totalitarian government. As bad as some of the influences in our society are, in this case the cure would be worse than the malady. But of course this is not the only alternative.

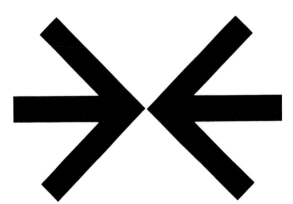

The debate between behaviorists and determinists cannot be resolved.

Both are right. Innumerable stimuli influence a man's actions.

At the dynamic local level in a cooperative territory, a group could band together and work in common to create its own orbit. It could strive to incorporate as many positive, rich, and vital elements as possible into its life-style. It could design the right kind of place for the right things to happen and produce such a lush environment that the young could not withstand the positive influence. It could truly determine its immediate world and could do it with those who shared common goals. After all, man is still evolving, and there is the possibility of a species beyond *Homo sapiens*.

We must accept what we are. In speaking of man, Shakespeare said: "how like a god"; Pavlov said: "how like a dog." They are both right. We have a biological inheritance we must respect. We need a private territory where the bonded pair can raise their offspring and a cooperative territory where orbit and scale are respected, where a local society is designed by its own inhabitants and has fit. Housing that respects all these elements will be successful housing.

A healthy neighborhood provides a control screen for the kinds of stimuli with which man must deal. Positive stimuli are encouraged.

CHAPTER TWO

Housing in America

*"There is no doubt about the sterilizing influences
of many modern housing developments, which, although
sanitary and efficient, are inimical to the full expression
of human potentialities. Many of these developments
are planned as if their only function was to provide
disposable cubicles for dispensable people."*

René Dubos

T

he urban family in a technological society is faced with inordinate difficulties. Modern life is too complex, too precarious for the average American family in the cities. Add to this the ill-conceived and inadequate housing that most city dwellers must endure and the correct mix for frustration and unhappiness is manifested. In the past, America provided opportunities for almost everyone who sincerely wanted a home and property to have them. The promise of property ownership is one of the reasons that people left their homes in Europe and other parts of the world to emigrate here, and the single-family residence was never considered a luxury, even in densely populated areas. A national posture infers that homeowners are more stable citizens, and in our hierarchy of status symbols, home ownership is a must.

As America expanded westward, the early settlers built their homes on land purchased from the government. These were simple, honest structures which solved the imme-

The sleeping loft in early houses was a
free and spontaneous space for children.
Its replacement — the typical child's
bedroom — is a less vital environment.

26

diate needs of shelter. Many of them had dirt floors, log walls, and lofts. Soon Americans were building more sophisticated homes that recall the architecture of Europe, but in a new and personal way. These early houses with great fireplaces, rough-hewn beams, magnificent plank floors, and simple furnishings were warm, excellent places to raise families.

As highly skilled craftsmen emerged in urban areas, residential building began to express a regional quality. The townhouses of Beacon Hill in Boston and Georgetown in the District of Columbia, the great manors of the South, the Cape Cod cottages of New England, the French Colonial mansions in New Orleans, the Georgian homes in Williamsburg, the wonderful old stone houses in Colorado and Utah, the great Victorian bric-a-brac mansions in San Francisco, the Spanish-style houses in Southern California, Arizona, and New Mexico — all in their own way solved the special needs of those who built them, and always the architecture achieved a certain nobility.

Houses in this period were more exciting and far above the average of what we find today, largely because a person simply could not go out and buy one. The individual

27

Early housing expressed regional qualities, such as the New England "salt box."

owner, striving to satisfy the needs of his family, had to assume the responsibility of creating the building; a dynamic relationship necessarily existed in which the owner and architect or builders pooled their requirements, ingenuity, and skills. Great architecture does not emerge without a concerned and demanding public.

In the 1870s our industrial centers began their most phenomenal growth. The problem of housing for factory workers was solved quickly and efficiently with virtually no humane considerations, and the housing business became a lucrative enterprise which bore a striking resemblance to warehousing. The shocking concept of "human storage" resulting in hideous slums has been the cause of untold misery since the industrial revolution, and modern society is still recoiling from its horrors.

The emerging middle-class *nouveau riche* built some rather attractive but often bizarre eclectic houses that had no relationship whatsoever to their own society or culture. Young architects today are taught that these buildings, because of their "bad" design, are the nadir of architectural achievement in the late nineteenth century. Yet if we look at them realistically, we see that they were imaginative pieces of work, despite their phony grandeur, and were likely very nice places for families. These are the houses that grandmothers own

The western pioneer home reflected the rugged simplicity of life.

today, the kinds of places to which kids love to return because they are filled with nooks and hiding spots that settle deep into childhood memories.

An interesting exception to the gingerbread house was the Prairie House, developed in the Midwest in the early 1900s by Frank Lloyd Wright. The brick bungalow with its substantial overhangs and hip roof represents a truly American architecture, logical and functional. The Prairie House indeed deserves its great influence. It established a good and healthy direction that was never fully exploited.

Just before World War I, a new concept of land speculation emerged. Developers went far into the countryside and bought large pieces of ground on which they built houses that were sold on a speculative basis. The subdivision was born — but without the anguish in which it exists today. An effort toward genuinely good planning resulted in such great achievements as Radburn, New Jersey, where forward-looking planners developed a community which could enjoy total separation of pedestrian and automobile traffic and great open spaces dedicated to recreation. All this happened in 1929.

Tragically, America disregarded this vital forward step, while Europe took it seriously, with the result that in the area of planned communities, we are years behind much of Western Europe. Housing with large areas left in landscaped, open space and with neighborhood centers has become national policy in Great Britain, Sweden, Denmark, Finland, and many other countries. As a result, these countries are in their fifth generation of community planning, while we are just beginning.

After World War II, subdivisions proliferated at an incredible rate, spilling over the countryside that surrounds urban centers. Everyone dreamed of finding the right kind of place to raise a family — a house on a piece of ground. There was accelerating need by the public and great profits to be made by the land developers. Immediate housing needs were filled, with the tandem profits made — and all achieved with badly designed houses in poorly planned subdivisions. Some areas became blighted, rundown, ugly, while others were treated more kindly with ivy and evergreens. But there was

29

something missing from the outset, something that has caused subdivisions to be dull and drab: There is precious little to do there! The residents are miles from parks, shopping centers, piano lessons, the theater and opera, the library and university, the swimming pool and football stadium. The houses are comfortable little boxes, located somewhere just off the highway leading into or from town, with no recreational facilities and no sense of community. In the postwar age of burgeoning resources and growing families, the men who produced and marketed these houses had returned to the monstrous concept of human storage.

And here we stand today. The typical American family has two bathrooms and a carport. The house has some technological conveniences that provide for the family's physical comfort; but few of these houses are part of a healthy, living community. We are presently experiencing another building boom, and the issue that must be faced is whether we are going to continue in the unfortunate pattern of the past. Most building today is being done in the same old way: A house on a street somewhere in the suburbs is the most one can hope for — and this is extraordinary since all the experts agree that this is totally wrong! There *are* new and exciting and healthy innovations. At the same time, there are still some bad, cheap, and shabby solutions. Because the primary function of a house is to provide a healthful place for the family, we must go beyond the traditional necessities of a warm space and a good roof. We must provide another necessity: the right kind of total environment.

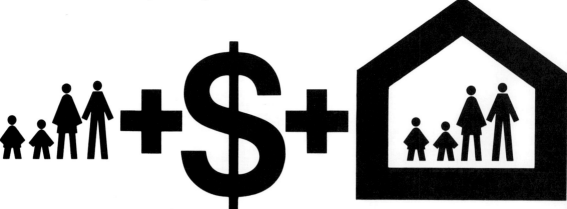

After World War II, people and money and housing nurtured a highly successful building industry; many fortunes were made.

What is total environment? It is the house as it relates to its own private yard and public open space as well as the neighborhood center. The house itself must go beyond shelter to something more adventurous and stimulating. We must not resort to superficial design trends such as Early American and Mediterranean, but must find honest and meaningful ways to express contemporary living.

There are those who are recommending mobile homes as a solution. And the people who should know better are listening. If America must resort to "disposable cubicles for dispensable people" or even accept them as a solution for the poor — if this is where free enterprise has taken us — then our nation's economic system has failed, for nations with a much lower per capita income house their people better. Other housing authorities are recommending high-rise structures. For the childless, the young marrieds, and the middle-aged or aged, these are adequate; but for family housing, they are deplorable. Children have a negative reaction to the crowded conditions and the anonymity that is forced upon them by high-rise buildings. It matters little whether identical units are placed side by side, as they are in many subdivisions, or stacked vertically in a high-rise — either way, they remain equally deadening.

The federal government feels a strong responsibility to provide for the indigent, and its agencies are sponsoring a variety of low-cost housing programs. Government officials seek good community planning as well as good housing in theory, but in their current desperation to meet increasing demands, they are in practice subsidizing housing projects

The final result was badly designed houses in poorly planned subdivisions where residents have precious little to do.

that have no future. A house on a street in an un-neighborhood is not the answer. Some specific plumbing, heating,
wiring, and structural requirements have been made manda-
tory. Why do these not include specifications for playgrounds,
open space, jogging paths, and village centers? The answer is
that this kind of requirement does not exist. We have building
inspectors but no public officials specifically concerned with
environment.

We have always had a kind of blind faith that new
housing would automatically generate a healthy environment.
With this inane attitude, we have built acre after acre of
pitched-roof boxes and then have professed surprise that the
result was the dull, barren wasteland which often charac-
terizes suburbia. Profits were made by the builders, and
property taxes were paid by the new owners whose children
still had to play baseball in the street. Their taxes paid for
parks too remote for their children to use. Twenty years ago
we knew that this was wrong, but since the system benefited
the housing industry and local government, the fraud has
been perpetuated — until now. Again it must be emphasized
that as a nation, we have had no vital philosophy of housing.

Excellent housing is not an impossible goal; there
are some well-designed new communities. Reston, Virginia,
just outside Washington, D.C., is an example of what a
planned village can be. It is built around an artificial lake
with a wooded area. A village square with a small church,
grocery store, restaurant, drugstore, and community center
surround a superb plaza decorated with sculpture that is
visually delightful and can also be used by children to play
in or on. A community focus is required for the well-planned
development, and certainly Reston is a most successful
example.

Another example is Columbia, Maryland, a newly
built city between the District of Columbia and Baltimore
that was planned to help absorb the burgeoning population
of both cities. During the initial planning stages, a variety of
social scientists was involved in determining the basic concepts of the new town. The result was the establishment of
a strong neighborhood pattern, with the grade school serving
also as a neighborhood center.

Westlake in Southern California is a beautiful community snuggled in the hills northwest of downtown Los Angeles. Constructed as a model community, it provides an extraordinary variety of recreational facilities and offers a smog-free version of what Southern California can truly be at its best.

All of these communities have important things in common: Shopping is nearby, and there are parks, year-round swimming pools, playgrounds, and community centers. There are places for a great variety of activities for children. The tot playgrounds and playing fields at Columbia, the Children's Art Gallery and play sculpture at Reston are excellent examples. Only a concerned community will provide a water sculpture that erupts every day at a certain time and where all the kids come running to enjoy the communal bath. Adults can enjoy golf courses, tennis courts, and a multitude of recreational facilities, as well as some exciting shops and restaurants. These communities thus far have been largely restricted to the upper middle class. Certainly they represent some extremely good experiments, and we can learn much from them. Yet the problem is to create high-quality solutions on a level that *all* Americans can afford them.

A columnist who recently returned from the Soviet Union commented on how incongruous it is for the Russians to be able to achieve fantastic technological advances in space exploration and at the same time be unable to provide running water in towns just fifteen miles outside Moscow. We must face the reality that the same enormous incongruities exist in America, and that while we too have achieved much in space, we have done a shabby job on the ground. Americans are not provided with good housing. Even two bathrooms and a carport are not nearly enough.

The private home is immensely important and has been from our beginnings. One reason for this is that a private house remains the best place in which to raise a family. Although this statement points out the self-evident, the realities of production and marketing of housing show that the family has not been well served, its needs have not been seriously considered. Providing an ideal environment in which to raise children has not been the primary goal of the real estate salesman, the building inspectors, the Federal Housing Authority, the architects — the housing industry as a whole. Each segment of the industry has performed its specific responsibility: If a house is safe, adequately designed, marketable, and meets minimum requirements, the industry has been satisfied. Yet these people, in their fragmented roles, are those who make the collective decisions. No one has been responsible for solving — or even formulating — the problem in its entirety. Why? Because we have simply never developed a unified, dynamic philosophy of good housing. Housing has just happened, and for most Americans a well-planned and -executed total environment remains a hope for the future.

A Healthy Family in a Healthy Community

When a family is a healthy family, it functions well in at least four specific areas: self-image, communication, rules, and linkage. The family collectively and all of its members feel good about themselves. Lines of communication are always open, and there are specific rules — democratically determined — which govern the social system of the family. The family is linked to institutions outside itself through a system of healthy relationships.

The same functions are necessary for a healthy community. The self-image of the community must be good; citizens must be proud to live there. Many channels — perhaps "excuses" would be a better word — must be provided for communication to occur among members of the community. There must be a system of rules — democratically determined — by which the social systems of the community are governed. And finally, strong ties must exist with the external institutions which affect the lives of citizens. These ties must include a direct relationship to the needs of community members, and there must be some means for representative expression.

This type of healthy community is not at all an unreasonable expectation in our society.

34

CHAPTER THREE
Application of the Theory

"Our capacities for sacrifice, for altruism, for sympathy, for trust, for responsibilities to other than self interest, for honesty, for charity, for friendship and love, for social amity and mutual interdependence have evolved just as surely as the flatness of our feet, the muscularity of our buttocks, and the enlargement of our brains — whether morality without territory is possible in man must remain as our final unanswerable question."

Robert Ardrey

We have followed man's progress from his primitive culture to his modern stance. We have discussed some well-developed biological and sociological theories now under study. We should therefore have considerable knowledge concerning human beings and their living patterns. From all this, a comprehensive philosophy of what comprises a healthy home environment should have emerged. But this has not yet occurred. Rather, in the absence of such a dynamic human environmental philosophy, some negative influences have stealthily infiltrated and taken control, and modern society — in a state of change — often substitutes the quick and tawdry for time-established, valid systems that have served well in the past.

If we are to redirect these negative influences, we must put forth a serious effort to determine the optimum kinds of activities that can occur in a healthy family situation. The important issue is to find the right kind of place for the right things to happen…and the solution must be the kind of housing that the *majority can afford*. If the elements of a healthy environment are missing, then regardless of its cost, a house can only qualify as human storage. On the other hand, if there are exciting spaces for rich, rewarding, even mind-expanding things to happen — given a reasonable effort by those who inhabit the space — they likely will happen. These are the kinds of elements that should form the basis for contemporary design.

There are few technological advances that contribute to a vital, stimulating environment, since the timeless elements we will discuss could have been incorporated in a house a hundred years ago. After all, as we discussed in Chapter 1, the house and yard comprise the private territory where the pair bond (man and wife) strive to do a good job of educating their offspring. The significance of territory cannot be overstated because it satisfies the most basic biological demand, and because only with their own private space can human beings achieve the sense of security, identity, and

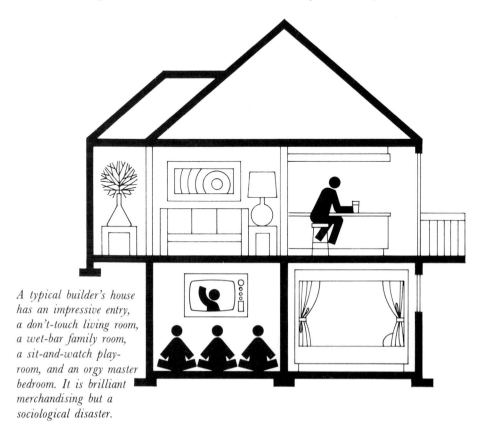

A typical builder's house has an impressive entry, a don't-touch living room, a wet-bar family room, a sit-and-watch play-room, and an orgy master bedroom. It is brilliant merchandising but a sociological disaster.

stimulation imperative for good mental health. And this, of course, refers to both the adult and the child.

A Place to Gather

Let us further investigate the necessary features of the private territory. Often there is too little communication between parent and child, especially during the adolescent years. One possible reason is that there is simply no natural place for spontaneous gathering to occur.

Many beautifully furnished living rooms do not inspire a family to gather together. But a fireplace with an inviting blaze is always enticing. Maybe it is the primitive, residual instinct, but most everyone finds it difficult not to be captivated by the warmth of a fire's flickering light. When fireplaces were the major method of heating a house, everyone enjoyed them. Perhaps not physically but certainly psycho-

A true family house offers a vital, imaginative, productive environment; it is a place for creative family living.

41

logically, they are just as important today. Where there is a fire, there is sure to be a gathering of people. Many large, spacious living rooms with magnificent furniture lack this one essential element. Many low-cost houses are built without fireplaces — yet they are as inexpensive to add at the time of construction as a carpet. This is not a design feature to be enjoyed only by the affluent. The hearth is the right kind of place for the right things to happen.

A conversation pit surrounding a fireplace can be an ideal family gathering area. No parent will have to recommend that the children join him in a properly designed pit. It will just happen. On the other hand, in Southern California, the conversation pit is being designed as a kind of orgy pit with a wet bar next to the fireplace. What a sick and distorted misuse of what was originally a very good concept! Yet this is a good example of negative influences creeping into home design simply for lack of a dynamic philosophy of the positive, healthy elements a house should provide.

A Place to Dine, Not Feed

Some twenty years ago someone came up with the idea of eating at a bar in the kitchen. This was thought to be an excellent dining solution for the family who were engaged in diverse activities. After observing kitchen bars for a generation, we must concede that they are extremely efficient; however, as they affect family relationships, they are less than desirable, to say the least. Dinnertime should be the one part of the day when the family gathers naturally. There can be no better time to enjoy each other's company. There is no more fundamental family ritual. Formal dining rooms are not necessary, but the dining area should be warm and inviting as a gathering place.

A Place to Play, or Just Relax

The family room is another concept of twenty or so years ago. The basic theory was good: The family room was to be a place for the family to congregate and for the children to play. The idea — though redundantly superimposing itself on the concept of the living room — negated the need for the "parlor," the "living" room. But even worse, the family room became a second-rate living room containing many "don't touch" articles that were out of bounds for children.

Forced from the living room, restricted in the family room — where now was space enough for the child to feel the full thrust of his energy? Living rooms should be family rooms, family rooms should be playrooms, playrooms should have virtually no rules other than for reasons of safety. What more could parents provide for children than a space for rough-and-tumble play, a place to paint or work on models and just be creative? If someone wants to paint a mural on the wall, what better decor could the room have? When one contemplates the variety of ways to create an exciting place in which to grow up, he finds that the possibilities are staggering: There should be ladders to climb, ropes to swing from, a platform to climb onto, a fireman's pole.

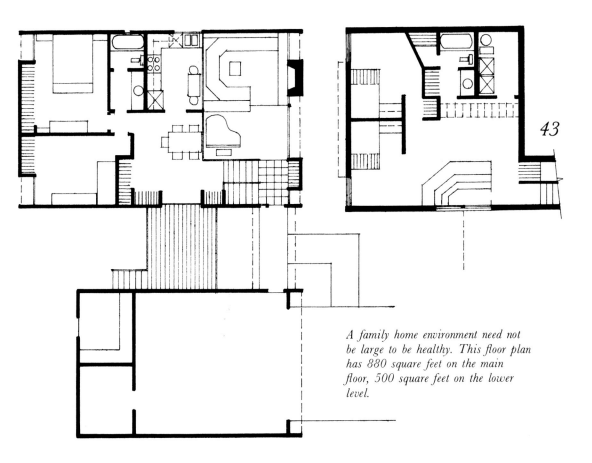

43

A family home environment need not be large to be healthy. This floor plan has 880 square feet on the main floor, 500 square feet on the lower level.

In a classic psychological experiment, a test group of mice was placed in a bare, white cage, and another test group in a colorful cage with buttons to push and levers to press — simple accoutrements to stimulate intelligence-building. Not surprisingly, the second test group in the vibrant, colorful environment developed a much higher level of intelligence than the first. It causes one to ask: How many children are brought up in the sterile, dull environment, and how few experience a vital, colorful, stimulating environment?

Behaviorists have stressed that advancement in behavior can be made through rewarding positive behavioral patterns; there are great possibilities in behavioral modification and behavioral engineering. The parent has an exceptional opportunity — as well as responsibility — to expose the child to a stimulating and at the same time secure environment. A conscious effort by thoughtful parents, or unconscious happenings supplied by television and friends, can provide the dominant behavioral base for the child. Here is an area of housing design that has not been thoroughly investigated.

Most of us can remember a kindergarten room where we first started our school experience, which had huge wooden blocks, or a large playhouse, or games, musical instruments — a truly vital kind of place. This was followed by the first-grade room from which most of the fun things had been removed, and suddenly school became serious and terribly dull.

We do the same thing with our children's environment. While they are very young, we buy them colorful and intriguing items to play with, from which they learn. But as they grow older, we lose our enthusiasm for educational toys at the very time when such toys could possibly accomplish the greatest good and be the most meaningful to the child.

We have discussed the trend — particularly in California — toward a wet bar in the living room and a family room of every upper middle-class home. The propriety of the bar itself is not in question; what is being questioned is its priority over an enormous number of other elements that could contribute so much more to the development of the family. Perhaps if the children were better cared for, the need for the bar would be less intense.

A Place to Be Alone

Each child, whether he has a separate room or not, should have an area that he can call his own. Each should have his own desk and dresser as well as shelves upon which he can begin his personal library and can display his hobbies. Also, a hobby closet or space — closed off from the remainder of the bedroom that can be left in a mess but which provides space for the child to follow a project through to the end without having to clean up — is imperative. This area need not be large; a small area will suffice to provide these necessary facilities at very little cost. If there are two children in a room, it is necessary to delineate "property lines" so that responsibility for caring for the separate areas can be assigned. And again, we must insist that such hobby areas — a place to be alone — should be available in low-cost housing.

The parents' bedroom should be a place for them to get away and relax — especially if there is a houseful of kids. If space is at a premium, it can also double as a study, or could also provide space for a hobby closet. There is a curious trend in America today toward making the master bedroom a rather flamboyant expression of sexuality with a king-size bed, wall-to-wall mirrors, a raised bed platform (a kind of sacrificial altar?), with lacy bed curtains, a sunken tub at one end of the room, and so forth. Aside from its being a rather gross form of decoration, there are probably few human beings who are equal to the challenge of the facilities.

A Place to Remember

We all remember the warm spaces that were a part of our lives during childhood: a window seat where we could sit and eat an apple and watch the rain or curl up with a book; a broad front porch with its swing, where young and old alike would gather on a hot summer's night; a magnificent staircase; a large fireplace with glowing embers; a two-story-high wall of books with a sliding ladder; a balcony area overlooking the living room; a stained glass window; a skylight with a shaft of light penetrating a dark space; a heavy, dark-beamed ceiling. Maybe it was just the way the sunlight filtered through a window and filled the room with a soft, warm glow.

All of these are elements that transform "just another place" into a special kind of place.

What will *our* children remember about the place that was their home? In some way these kinds of things must be made to happen in low-cost housing. Certainly expensive elements will be impossible, but something must be added. Merely dividing nine hundred square feet into three bedrooms, a bath, a living room, and a kitchen-dinette will not create a memorable space for living.

A Healthy Neighborhood or Cooperative Territory

What is a healthy neighborhood? To be sure, there is an infinite variety of healthy neighborhoods, but many of them will have these elements in common: First, there is a common space for children to congregate. In nature, the highest function of the cooperative territory is guarding and protecting the young, as well as educating them for survival.

A healthy neighborhood includes a house, plus a yard, plus open space, plus a village center; it is a cooperative territory.

The younger children should be able to gather in small play areas, with play equipment appropriate to their ages, within close proximity to their homes so that they can be supervised by the parents.

Older children need larger playing fields, bicycle paths, streams to wade in, woods to investigate. Ideally, they should be able to get to school without coming into contact with heavy traffic arteries. Their parents should have some method — or excuse — through school, church, shopping, or other community activities to get to know each other. Cooperative territory must provide the same elements we demand of individual territory: security, stimulation, and identity. And again we must emphasize that it is here that perhaps the most significant education of the young occurs.

In a neighborhood, there should be a small commercial area where neighborhood service shops can serve the basic everyday needs. This element alone can give an area a neighborhood identity. Small service areas just naturally happened before the automobile became prevalent, and the people in the area naturally responded to this neighborhood focus. Large shopping centers have been a major factor in destroying this possibility; because it is more efficient merchandising to establish larger centers does not mean that it is sociologically beneficial. Again, this is a case of negative elements being introduced for lack of a well-defined philosophy of how things could and should be. This is not to say that healthy neighborhoods do not exist in America; but it is tragic that there are so few.

What we have been discussing here is not a theoretical planner's dream but has already become a reality in many of the new towns of Europe where housing is planned around neighborhood recreational facilities as well as neighborhood centers; and this is done on a grand scale with great numbers of people enjoying the tremendous sociological benefits of this kind of living.

We must emphasize that a true standard of living should not be determined by the number of things a person has accumulated, but rather by the immediate variety of possibilities for individual fulfillment which exist in his environment. A thirty- to fifty-acre subdivision with no parks or common areas is organized to benefit the land developer, the builder, and financial institutions, but is reprehensible as an environment. A golf course or lake or any other single-function open space is so limited in its use that it also fails as an environment.

We would be amazed at the number of facilities such as parks, playgrounds, enclosed swimming pools, and so forth, that a community of fifteen hundred people could easily afford. We try desperately to keep local taxes at a minimum, then we pay ten times as much for amenities such as golf courses and swimming pools because we are forced to join country clubs that are miles from where we live.

As we must not separate the house from the recreation areas, neither can we isolate house and open space from the neighborhood, nor can we separate the healthy neighborhood from a vital community. To have a sense of place and a sense of belonging to that place, and to have a feeling of common goals and a cooperative spirit with one's fellow humans are imperative for good mental health. It is also necessary to know and feel that you have some say in the way things are done rather than feeling that you are ruled by a remote, anonymous, and arbitrary government. A com-

48

The Expanding Parameters of Environment

For the apartment dweller the hallway is the beginning of a hostile world.

A private yard provides an expanded parameter of influence for the individual.

Common open space expands the parameter and infuses the feeling of community.

munity with its own identity is the right kind of place for the right things to happen.

John Kenneth Galbraith in his book *The Affluent Society* describes a family leaving the luxury of their air-conditioned house, getting into their air-conditioned car, and driving miles to a picnic area strewn with beer cans and refuse. The contrast between their private wealth and the public poverty that must be endured is overwhelming. But let us follow this hypothetical family just a little farther; after their picnic they drive down a street of strip development that achieves an incredible level of banality, and finally arrive at a supermarket. One of the family members goes inside and asks a clerk whom he has never seen before, and will probably never see again, where a certain item can be found. The clerk replies, "Row 10, Section H." And as the shopper's eyes follow along the barren ceiling to the various designated rows, he strikes out across the great unfamiliar and lonely space, bumping carts — and defensive eyes — with people he has never seen before and will never see again. The lack of social linkage — the age-old pattern of developing strong associations with people you serve and who serve you — combined with public poverty together negate any possibility for community to exist.

49

A *village center provides the focus for community activity which is truly participatory.*

Food gathering capability or economic capability can be added by providing a co-op farm or a co-op factory.

*In early communities the individual had power
to participate in the decisions of the community.*

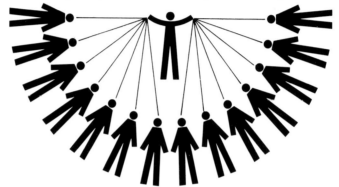

*As the community deteriorated, centralization of power left the individual
powerless, subject to arbitrary decisions from afar.*

50

*Power can be equalized if communities are re-created to provide for individual
input through the community to the remote power structure.*

Most Americans live in houses with many of the luxuries of a private yard (private territory), yet they are forced to share their open space (cooperative territory) with too many people, too far away from where they live. In cities where densities become too severe, we find the characteristics of a behavioral sink. A person's alienation is further reinforced because he does not *belong* anywhere and is forced to associate with everyone — therefore no one! He shops and finds his entertainment somewhere along the road. He lives a partially settled, partially transient life — a life with an ever-changing orbit.

Jane Jacobs, an eminent architectural critic, feels that planning can have some absolutely deadening effects, which it undoubtedly can. However, she places a kind of abstract hope in the right things happening spontaneously. We can assume the mathematical probability that a good environment will occur spontaneously about 50 percent of the time. Probability is fallacious here, of course, because good environment simply never has "happened" 50 percent of the time and never will.

An unfair advantage is assumed by self-centered commercial interests that are the antithesis of good environment, yet these interests are constantly engaged in planning our environment for their profit. Until a sound philosophy with established priorities of what comprises a good total housing environment is established, we will continue to find ourselves entirely at the mercy of these inhumane forces. What is good for human beings is the question that *must* precede today's icon, "Is it good for business?"

In a complex modern society, where television and the children's peers are becoming the strongest influences in a child's life, it would seem imperative to search out those elements that can reinforce the parent-child relationship. Moreover, with all the activities that are available outside the home, the parent can miss the deep satisfactions gained only through a very real dedication and commitment to the children. Considering all the ways that man finds his deepest satisfactions in living, just possibly the parent-child-family relationship can be the most gratifying. Many people are convinced of this, yet this idea has not provided a vital direction in contemporary life. If anything, the opposite is true. Residential architecture and community planning, then, have no higher purpose than to promote creative family living.

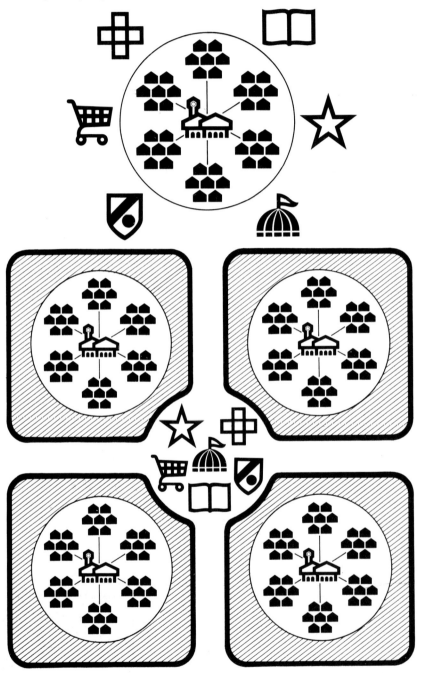

52

The ideal neighborhood provides private territory and co-op territory. It provides security, identity, stimulation, and participation for its members through a neighborhood center. Institutions are outside the neighborhood and serve to relate one neighborhood to another.

CHAPTER FOUR

Residential Architecture

Ⅰn certain tribes in Africa the woman is the architect, builder, and decorator of the home. The houses are built of clay which has been molded into wonderful sculptural shapes and left to dry in the sun. Later they are whitewashed, then decorated with magnificent wall frescoes. Here are houses that fulfill the idea of residential architecture for they are honest, regional in character, and are an authentic, intuitive expression of a way of life. Could there be a more significant environment in which to raise a family?

A low-cost housing project in Puerto Rico originally consisted of many identical little bungalows. This condition was short lived, for the new owners, dissatisfied with what was offered, began painting huge designs across the façades. An extraordinary variety of colors and added paneled shutters with magnificent painted shapes so totally transformed a very ordinary development that it was hardly recognizable. These people, "primitive" according to our standards, were unwilling to accept the status quo. It was imperative for them to establish a personal identity for their new home.

A young couple in Southern California was making the rounds of several housing projects, trying to make a decision on the purchase of their new home. They were not surprised that the houses built by the various builders had exteriors that were similar, with floor plans that were almost identical. True, they had their choice of three exteriors, French Provincial, España Mediterranean, and Early American. Since they had an Italian Provincial living room set, an Early American dinette set, and an Oriental Modern bedroom set, they thought it only fitting and proper that they choose the España Mediterranean. Twenty years ago, the couple was different and the exterior names were different, but the problem was the same.

We have stressed the enormous biological and psychological significance of territory, and certainly the house is the most immediate personal expression of territoriality. To foist grotesque, trivial "architecture" on the public verges on criminality, for territory must provide security, *identity*, and stimulation. Shallow, meaningless shelter can only encourage anxiety, conformity, and boredom.

North of San Francisco along the coast highway is a small community called Sea Ranch. Here simple, honest, unpainted structures cling to cliffs that fall precipitously to the sea. The community is new and all the houses are new, yet there is a timeless quality about the place. There is a rough, unpainted old barn, like many similar structures along the coast, that really told the architects what to do. The grim, windswept yet magnificent setting posed some definite problems. Salt spray from the ocean, fierce winds, magnificent views, and an austere, grassy knoll for a setting were all design problems that had to be resolved. This was a special kind of place that required the same honest, intuitive approach that primitive men have always used in determining the shape and form of their shelter. What was built here could have been done one hundred years ago, yet it is an honest expression of the current life-style of the Bay Area.

Each part of America has its special quality and has a specific set of problems to be solved, and from this a regional architecture should emerge. A good way to discover some valid and logical solutions is to look back and see how the early settlers in a region solved the immediate needs of shelter before a multitude of external, arbitrary influences appeared. Another way is to look about and try to determine what materials and design would blend best with the natural setting.

The history, the culture, the mix of the population of an area provide distinctive characteristics worthy of expression. The buyer of a home must ask himself if a design element is appropriate. Does it have "fit"? A regional architecture is worthy of expression, particularly in a rootless society desperately in search of an identity. Stucco, tile roofs, the Colonial Spanish influence interpreted in a contemporary manner are some of the unique qualities of a residential trend emerging in Southern California. The Bay Area has its own long-established style and its own method of building, which together have produced some of the most exciting houses in America.

Architecture in the Northwest has developed its own vernacular, and again there are some extremely handsome examples. The mountain houses of the Intermountain West are just beginning to emerge. Several architects have started an interesting trend in the Southwest with some exciting possibilities. The Prairie House of another era comprises the only valid trend that has emerged in the Midwest, while the East and South are still generally hung up on the past.

This has been an observation in very general terms, and certainly there is an extraordinary number of exceptions. There have been excellent contemporary houses built in many parts of the country, but they have not been an expression of their particular areas — nor were they meant to be — and the regional architectural styles we have mentioned have certainly not had total influence in their specific areas. The aluminum-clad builder's box has had its destructive influence in almost every part of the country.

It would be wonderful to find the infinite variety in building that we find in our natural landscape. America would be a more exciting place with various parts of the country expressing their very special characteristics. And who knows, in fifty years Europeans might find America to be a great tourist attraction because of its picturesque villages and hamlets!

Structure

The structure of a house is a simple engineering problem requiring short spans for the structural members. An honest structural expression has always been a necessary ingredient to good architecture. Since huge spans are not required, the best use of the structure is to provide possibilities for exciting spaces and some nice detailing to occur. In the restoration of Williamsburg, Virginia, it is extremely enlightening to see the simple, direct, honest yet magnificent way that colonial houses were built. But it comes as a shock to realize that we are building virtually the same today. Because the problem remains the same, a modern solution need not be revolutionary.

Young architects secretly yearning to do large buildings occasionally yield to the temptation of using a structural tour de force such as a diamaxium dome, a hyperbolic paraboloid, a cable suspension system, space frames, and so forth, for residences. The result is sometimes an exciting structure, but almost always a bad residence, simply because a technique developed specifically for large open spaces is being used for a house, which is naturally composed of many smaller, more intimate spaces. Builders, on the other hand, use prefabricated wooden trusses which are a fairly economical but extremely dull structural element. Trusses can be very handsome, however, if they are exposed. Post and beam construction can also be used effectively and economically and provides many possibilities for vibrant spaces.

It should be noted that high-rise building is not being discussed. The author feels very strongly that high-rise housing has no validity sociologically, even in cities that have achieved extreme levels of population density. To encourage this inadequate kind of housing in overpopulated areas would be wrong.

During World War II the citizens of London decided they had already surpassed a healthy population density. For that reason, after the war they discouraged more building in the city and specifically encouraged housing in new, planned cities in the countryside. The new housing on less expensive land consisted primarily of townhouses with minimal front and rear yards, along with necessary, planned recreational amenities. They were simple, economical structures that the majority could afford. The point is that a high-rise is simply no place to raise a family. To further compound the socio-logical error of choosing to build such housing, the construc-tion cost for high-rise building — including the kind of built-in forced conformity that we would not endure in buildings on the ground — still remains almost twice that of wood frame construction!

With proper planning, even where high densities exist, a single-family dwelling is not only possible but remains the correct solution.

Prefabrication in housing seems to be inevitable, although the same old method of stick building has remained reasonably economical. There is a great deal of excitement about new methods in building techniques, and certainly technological breakthroughs will occur, but we have thus far not achieved anywhere near what we had hoped. An officer in a large housing company maintains that prefabricated homes, using prebuilt wall and roof sections erected on the site, are definitely more economical to build. The entirely prefabri-cated, modular house, comprised of twelve-foot-wide sections completed at the factory and then moved to the site, as yet has not been able to compete in terms of cost.

63

What does all this mean? Simply that some form of prefabrication is inevitable, and it is undoubtedly the way of the future; but we still have a long way to go, and there are no quick, easy, universal solutions. The great contribution of prefabrication has not been the system itself, but that it is forcing the home-building industry into more organized systems of production. The criticism leveled at the housing industry (Chapter 7) for its lack of technological capability is misplaced, for its technology is well ahead of the antiquated financial systems and government regulatory systems that actually control the end result. Even our concepts in community planning deserve much more criticism than housing technology. We must also remember that home building is still an area in which there is a variety of craftsmen; and although their quality of work has deteriorated, they have not yet been superseded by technological processes. True, there is social value in having a great number of people who can still create things with their hands. Do we really want to change all this?

Mobile homes are hardly worthy of mention, except to say that some people who should know better are seriously considering them as a solution for low-cost housing. Structurally, they are minimal in every way, and in some instances have become a hazardous firetrap. Economically, they are a failure because people who buy them must pay higher interest for minimal shelter that depreciates to a level of nearly nothing about the time that the mortgage is paid off. America's infatuation with built-in obsolescence has been a curse in the automobile and small appliances industries, but to resort to this kind of thing in housing is a grave mistake. To encourage temporary housing in an already rootless society would prove in the long run to be a sociological catastrophe. Owning your own shelter but renting the territory on which it is located is as bad as renting in high-rise housing. Exciting, clever, and unique structural systems can make a great contribution to architecture. However, in housing it is imperative that the system be both appropriate and economical so that enough remains for solving the more significant functional needs.

The only positive contribution that the mobile home industry has made to home building is in manufacturing a product that is habitable at the time of sale; built-in beds, carpeting, and draperies are part of the package. It generally takes a huge added investment to make conventional housing habitable. In the California market, and for that matter in many other areas, houses are not even provided with enough lighting for one to safely walk from one end of the house to the other. Lamps, carpets, draperies, and furniture must all be purchased. The customer acceptance of the immediately habitable mobile home should prove that this would be an excellent direction for the conventional housing industry to follow, and built-ins designed for critical human needs have immense possibilities.

Function

The functions of the parts of the house can be divided into two general categories: (1) formal-adult-passive and (2) informal-child-active. The entry, living room, formal dining room, and master bedroom make up the formal-adult area, while the kitchen, dinette, family room, playroom, children's bedrooms, utility room, and mud room comprise the informal-children's area. Almost always the formal-adult areas are situated and well equipped, and the informal-children's areas are stuffed into the leftover space. Yet the informal-child-active areas are the spaces where perhaps the most important things happen. This is of course based on the premise that raising a family is the highest priority activity in the home and that the family's everyday use thus supersedes special activities such as entertaining.

The living room requires no larger space than a place around the hearth and room to play parlor games and listen to music. The formal dining room deserves only the amount of space that the frequency of family use justifies. The master bedroom should be pleasant and spacious, but a space well used need not be large to achieve a luxurious setting.

Generally speaking, the formal area should not exceed one-third of the floor area for the simple reason that its activities cannot justify it. What is of greatest importance is the *quality* of the formal area, not the quantity. A place to gather around the hearth, enhanced by a noble fireplace, a beautiful setting for a grand piano, a place for books, richly textured walls where paintings can be displayed, a warm, cheerful, inviting space, creates the right kind of atmosphere for the right things to happen.

The informal-children's area should inspire activities of all kinds. It matters little where they happen or how, only that the possibility is provided. If there is no place where

65

painting is allowed or encouraged, children probably will not develop an interest in art. If there is no workbench, children probably will not be mechanically skilled. If there are no books readily available because there are no bookshelves in the children's rooms, there will be no enjoyment of reading. If there are no rough-and-tumble areas, the kids will probably be off to the neighbors who do allow it. Finally, if the play-room is not really a place to play and if the kids' rooms are not comfortable, exciting places to be, and if as a result the kids are mesmerized by television, the house can be a failure, inhabited by unhappy children and neurotic parents.

The current phenomenon of large numbers of young people leaving home during late adolescense is indeed puzzling to older generations. This is not to suggest that there are some very simple, all-encompassing reasons for this occurrence. If there is little communication between child and parent, and if there is no vital life-style worthy of emulation, and if all that is being forsaken is a barren, humdrum existence, then why *not* leave? If raising children is not a warm and wonderful adventure for the parents, what can the children possibly feel? Again, a house itself cannot make the difference. *People* make the difference. But a house can be a powerful expression of a meaningful way of life, and it must provide the space for good things to happen.

The behaviorist school of psychology maintains that a particular setting influences all the individuals in it. Behaviorists further maintain that our genetic inheritance is less important in predicting behavior than the physical environment. If this is indeed true, the activities and possibilities that a house provides take on enormous significance.

Often in housing periodicals children's rooms are shown in which the space functions both as a quiet sleeping and study area and as a playroom through the use of sliding partitions that open into the adjoining bedroom. This is an excellent idea — not new, but very effective. How seldom these concepts are applied in mass housing! The bedroom in a smaller house should also function as a hobby area. A small (four-foot by four-foot) closet placed perhaps under a bunk bed can provide an excellent space where messy projects can be carried out but not at the expense of a messy room.

The matter of the direct function of a room will generally be readily solved. It is only when we go beyond the immediate purpose that we discover options that are possibly more significant. For instance, the utility room can be a little larger and function as an art studio as well as a great hobby area. The playroom could open onto the lower floor or upper floor, providing a two-story space with balconies, "slippery slides," ladders, firemen's poles, wooden planks, a backboard for basketball, and it could be decorated with supergraphics. A wonderful and mind-expanding kind of space is thus created. The one parental criticism would be the possible danger of doing things so high off the ground. The rebuttal is that children do all these things outside, so why not let them happen indoors on a padded carpet?

Another imaginative possibility is a crow's nest over a closet that has its own skylight with colored glass so that the kids can retire to their private hideout and see how different the outside world looks through red, blue, and green glass. More than one way of entering a bedroom — perhaps a hinged panel behind a bed or desk — makes the house much more functional for a great variety of games. A house that functions well for hide-and-seek is undoubtedly more successful than one that functions well for a cocktail party.

It can be a rich and rewarding experience to raise a family in a vital environment that provides security, identity, and stimulation — to watch children pick and choose those things that they do best and that will help to mold them into what they become. Here is the great opportunity for parent-directed behavioral engineering. The architectural design can and must create space for these great possibilities.

Design

Too much of recent housing "design" cannot qualify esthetically as architecture. The simple little box with holes punched for windows does not have enough happening on the exterior for one to presume there is any design whatsoever. The cozy Cape Cod cottage snuggled into the trees is an inviting symbol of what a home ideally could be. The shiny aluminum mobile home snuggled into a grouping of other mobile homes achieves only a symbolic void. The external appearance of a house should somehow inspire the feeling that here is a good place, a beautiful place, a proud place to return to. If contemporary housing cannot provide a symbol that can satisfy our emotional attainment, then it has failed.

Several years ago a well-designed, prefabricated, all-aluminum house was put on the market. It was a financial failure. Externally it resembled a small office building, clinic, or store...at least this is what the buying public decided. Cries that people are backward and that they just do not know what is good were heard from many professionals. The truth is that the house did not fulfill people's ideas of what a house should be, and who is to say they were wrong? The all-aluminum house was a strong expression of an advanced technology in housing, but it did not convey the feeling that it was solving human needs.

Architecturally we have broken with the past, but we have not destroyed it. The Cape Cod cottage can never have the meaning for us that it did for those who built it originally. But we must find our own idiom that more effectively portrays our present way of life, and as we develop a more vital life-style, certainly an exciting expression in architecture will emerge.

68

CHAPTER FIVE

Decorating and Landscaping

"As change accelerates and complexities multiply, we can expect to see further extensions of the principle of disposability for the curtailment of man's relationship with things."

Alvin Toffler

T homas Jefferson's contributions as a great political philosopher have had great impact on our way of life, yet his contribution as an architect has gone relatively unnoticed. In a period of powdered wigs and excessive use of perfumes, people were trying desperately to rise above being human. Some of the furniture design and general interior decorating that emerged from this period express an extremely shallow, cosmetic, superficial way of life. In many ways it resembles our own age with its over-emphasis on the right soap, deodorant, toothpaste — all of which will take us beyond just being human. For that reason Monticello (Jefferson's home), with its multitude of creative innovations, totally disregarding the respected fashion of its time, speaks out to us now and rejects the hollow fashion of the moment. Monticello is not just a house, it is a relevant way of life, and the way in which it expresses a simple, honest, intelligent, inquisitive life-style achieves an enormous significance for us today.

A home's decor should reflect those who inhabit the space and the quality of their daily living. The chairs, tables, and art objects — all the items added to the architectural space — should have a meaning to them personally. Ideally, the owner should know the artists responsible for the art objects he buys and should have some knowledge concerning his furniture and the philosophy of its design. If properly done, the home should be the owner's special kind of place.

Unfortunately, a room can be well designed but still remain a drab, uninteresting, and totally anonymous space. A possible reason is that the room is simply an interior decorator's product — the owner was not involved enough during the decision-making process. As a result, the space resembles a furniture showroom. The furniture choice and decor preceded the determination of the kinds of activities that should take place there. There are no magazines to read, no stereo, no piano, no games. There is probably no art except for some anonymous prints, and as a result the room fails on two counts: First, there is virtually nothing to do there except sit (activities available in a room are the elements that give it life). Second, the room does not give the slightest hint of those who live there. And one must assume that the owner has achieved nothing and his life is fulfilled only as a consumer. An anonymous room is as dull as a person with no interests. Without doubt, this is the kind of room that will seldom be used.

An exciting space was created in the living room of a sculptor friend. The walls, bookshelves, floor, and tables are replete with art objects. One does not notice what kind of furniture, draperies, and carpet the room has because this amazing display of art is overwhelming. The house was a typical builder's bungalow that was fairly well done but lacked any special quality. Yet the owner had so transformed a normal kind of space that it has a distinct character. The room achieved a superb identity, for it mirrored the interests of the owner. Could a room possibly express anything more significant?

Another young couple created an equally amazing living space on a small budget. The couch is made of an old packing crate, as are the bookshelves, coffee table, and so forth. The light fixtures, bare light bulbs with peach cans painted black, are used as directional lighting. Sketches, serigraphs, and signed prints are the only expensive objects in the room, along with some very fine pottery. There are no mass-produced objects anywhere. The design is theirs, the approach is theirs, the room is theirs.

Another unforgettable experience was a visit to a Maybeck house (designed by the very famous architect in turn-of-the-century San Francisco). The interior was built of natural redwood. The living room had only a grand piano, an Indian rug, some magazines stacked in a corner, and a few paintings, but the room was magnificent. It was apparent that the woman of the house was more student than housekeeper, but the space required little effort. A huge precast, exposed, concrete fireplace and all the natural wood superbly sustained the room. Any additional effort to decorate would have been most unfortunate.

A *finished* house needs no decorating in terms of cosmetic applications of wall surface materials such as wallpaper or prefinished paneling. In fact, when such second-rate materials must be resorted to, the house has already failed.

Presently, in Southern California builders are notorious for providing unfinished houses. The home buyer is forced to purchase a white gypsum-board space that requires various kinds of wall finishes, light fixtures, and so forth, before it becomes habitable. This of course is a bonanza for the interior decorators, who hang yellow flowered wallpaper all over the place and specify gaudy cut-glass chandeliers in conspicuous areas. Such gimmicks do lend a little sparkle, but at a tremendous expense to the owner. Good architecture, completed architecture needs no decorating of this kind. If the architect and the builder had not abdicated their responsibility, the often quick and shallow efforts of the decorator would not have been necessary.

Many people say that they like to change their interiors because they tire of them and therefore would not like such final finishes as wood and brick. The reason they need to change their rooms so often is that they are surrounded by shallow, artificial materials that become wearisome over extended periods of time. A good architectural interior with wood paneling, beamed ceilings, and brick floors needs a minimum of decorating. A variety of paintings, sculpture, and pottery with simple, classic furniture creates the kind of quiet dignity that wears well. The owners of such a house are collecting timeless items that through the years will increase in value and, more importantly, will provide a cultural environment that their children will slowly learn to appreciate. As the parents grow old and the children form their own families, these things become heirlooms. What a marvelous thing it is to establish a quality of living that the children are committed to carry on in their own way with their own families.

How many people will spend considerable money on a highly decorated picture frame for a print or a very cheap painting when they could get an extremely good painting with a modest frame for approximately the same price? It is amazing how people will buy mass-produced pottery that loses its value the instant it is purchased when they could get an original from a local potter for often the same price. These are the people who raise their children in a cultural void.

A grand piano costs about the same as a new automobile, but it retains its value much longer. Here is an item that can utterly transform a room. Of course, it is ridiculous to have a piano that isn't played, but it is an elegant, timeless element that adds profound richness to a space.

Good decorating should solve the problems of everyday needs first. The entertaining carried on by adults, even if they have an extremely active social life, will not account for very frequent use of a space. Too many living rooms are designed for cocktail parties, too many dining rooms are designed for formal dinners, and too many family rooms are designed for entertaining friends rather than living with children. This is the subtle way some parents force their children out of the house. If the parents are committed to doing a good job of raising their children, they should allow this to become a significant basis upon which design decisions are made. Such

an attitude could develop into an exciting and truly meaningful trend in decorating. Good design must have a sound,
logical basis. Certainly there could be no more significant
design philosophy than trying to give more life and vitality
to our daily living.

Many homeowners not only disregard their children's needs, but also their own. Their concern about where
guests will hang their coats and where they will go to the
bathroom sometimes achieves almost psychotic proportions,
and as a result everyday needs are subordinated. One client
revealed verbally what many think but would never dare say.
With reference to a master bedroom area she commented,
"People will like it." There is an extraordinary amount of
decorating based on what *others* would like and be impressed
by instead of the needs of those who inhabit the space.

The parlor is a revealing development in housing
design in California. It is simply the result of decorating based
on the false premise that a living room should be a lovely,
clean, nothing-out-of-place kind of space that should primarily be impressive. The decorators have done such a good
job of it that now no one dares use the parlor except when
guests come. More practical people felt that if they were not
going to use the damn room anyway it might as well be only
half the size, so we have the emergence of the parlor, an
almost totally wasted space for everyday living.

Interior decorators have had a tremendous influence
on housing design, and it might be added that this influence
has been largely undeserved. Most decorators are paid a percentage of the cost of the materials and furnishings they provide. They therefore have a vested interest in the kinds of
furnishings they supply, which should negate their right to
decide. Certainly there are significant exceptions, but the
decorating field as a whole should be held responsible for the
dismal quality of decorating in America. Why is it that the
typical European furniture store provides an unbelievable
variety of well-designed, contemporary furniture, fabrics, and
pottery — at prices that a great many can afford — while its
counterpart in America provides a dreary collection of Early
American, Italian Provincial, and Mediterranean furniture,

along with a host of mass-produced trivia in questionable taste. (In all fairness, there is the exceptional furniture store that carries well-designed items, but almost always they are very expensive. Good design does not have to be expensive. In fact, the main premise of modern design was to reject the traditional design of the past because its complexity made it expensive. What a strange paradox, then, to have bastardized traditional flooding the cheap market!)

The difference in the European and American approaches is a rather complex question, but one in need of an answer. First, furniture designers are possibly at fault for the simple reason that too many have designed furniture around the slick school of modern design. Stainless steel is not everyone's bag. Good designers have gone after the commercial items such as desks because that's where the big money is. Those who have designed in wood have simply made it too expensive. In order for the average person to get the warm, woodsy look, he has to resort to poorly designed traditional. Decorators were committed to contemporary design for awhile, but they simply gave up, or "sold out." Scandinavian designers and decorators knew that they were creating a superior product for contemporary living and they were therefore willing to put forth the time and effort to educate the buyer. How unfortunate that the American decorators took the easy road and put profits before purposes. How unfortunate also that they could find designers and manufacturers to supply their shallow demands.

Through the years decorators have jumped from one direction to another and consequently there has been little agreement between decorators and architects. But worst of all, they have continually designed spaces that could not endure human habitation. They simply could not be concerned with the needs of eating, breathing, sweating, eliminating *Homo sapiens* who are just not that immaculate.

This brings us to the question of whether or not we really need decorators. After all, we do not use specialists for picking our clothing. What we do need is a real commitment on the part of the owner to find the kinds of things that are best suited to his way of living. Interior decorators, void of any sound design direction, should remain in the area of commercial design for commercial interiors, where the product is generally well designed and where decorating has less sociological significance.

Our real hope lies in the furniture designers and manufacturers who have the real power to change things. A well-designed, inexpensive line of furniture based on creative family living remains an urgent need. Two kids and a dog carrying on life in a natural way would utterly pulverize the kinds of furniture which designers and manufacturers so often promote. In too many instances such furniture does not encourage the use of space but the abuse of it.

Regional Design

Another element worth mentioning is the regional approach to decorating. Since each area has its special characteristics, mix of population, and unique sporting and other activities, a regional style of living emerges that should find expression in decorating. Mexican folk art in Southern California, Indian rugs and pottery in the Southwest are good examples of elements that can help establish a direction. The Bay Area, with its hardwood floors, imported rugs, and many other imported objects, with particular emphasis on the Orient, thus expresses its role as an internationally famous seaport. New England on the other hand has a rich colonial heritage, and the heavy use of antiques mixed with contemporary provides a rich and appropriate regional direction. Elements that draw heavily from the past yet are not completely dominated by it are the basis for the design philosophy. Such combinations make allowances for the commendable desire on the part of most people to create interiors for their homes which are warm and comfortable but also colorful and up to date. It makes provision for the stability of the past and the excitement of the present to exist in our homes as they do in our lives.

The slick and sterile school of decorating is proving highly successful for offices but has never sold as a concept for home interiors, despite many very excellent examples of the design. People somehow feel more at home in the woodsy warmth of a Mediterranean style. Why can we not develop a warm, regional approach to design that can appeal to many people and that can establish a direction around which potters, weavers, sculptors, and artists can direct their efforts? The answer, of course, is that we can, and in some localities, we already have. On the other hand, the Mediterranean influence in decorating does not really make a lot of sense anywhere in America. At least the Early American trend, as phony and shallow as it was, had roots in our own past.

Decorating for Low-Cost Housing

Low-cost housing should have a considerable amount of sturdy built-in furniture so that the buyer can find the space livable from the outset. Almost everyone who purchases a house extends himself a little beyond his limits, and this is particularly true in the case of low-cost housing. If the builder simply offers a hollow shell, the house may be lived in for many years with minimum furnishings, and the quality of living is little improved over the previous housing. Built-in beds, dressers, and chests in the children's rooms, as well as in the parents' room, and a built-in couch in the living room would solve basic needs.

The opportunity to start anew with an expanded view of living is one of the main purposes for providing low-cost and subsidized housing. And of course an expanded view is unthinkable in an empty space. This concept can be criticized on the basis of increased building costs, which are very real. But at the same time, providing anything less than a house that is livable is not providing housing at all. We must find a way — and a simple way is to provide houses with built-in furniture, a higher appraisal value, and therefore a higher loan value. Another objection is that every interior would be furnished alike. This again is true; however, furniture is fairly insignificant if creative people give emphasis to other decorating items. Even if this does not occur, we can rationalize that a reasonably high quality of living can and will occur in similar well-designed spaces.

Good decorating can transform an ordinary space into something very special, but its real significance will be determined by how the possibilities for more creative living are provided for the family and how well the decor expresses who and what they are.

Landscaping

A living space with no private yard is too minimal a solution even to be considered valid housing. Many high-rise buildings have excellent playgrounds nearby, but a very important element — the private yard — is missing. On the other hand, a house with a generous yard but with no common open space in the immediate area is also too minimal a solution. We will therefore discuss the private yard as well as public open space as established minimum requirements.

We have discussed our biological need for territory, and certainly the private yard is its fulfillment. We have also discussed our need for community or cooperative territory, and the public open space satisfies this demand. Private yard plus public open space are established as minimum requirements because they fulfill definite biological needs reinforced in our species through thousands of years of evolution.

Architecturally we generally make the greatest investment in the front of the house, and this is equally true in landscaping. The large front yard is not unique to America, but it is our standard solution, which is not the case in many other countries. The front yard provides an excellent setting for a residence and, visually, a beautiful neighborhood. If it is judged in terms of an outdoor living space, its merits are definitely reduced; however, if walls, fences, and gates can be used to create private areas, the front yard can be made much more functional and often not to the detriment, but rather to the added interest of the overall street design.

The front yard is an excellent place to provide a formal garden, outdoor living area. Earlier we mentioned that a living room which provides nothing to do is really a very dull room. The same can be said for landscaping. Landscaping that is most interesting provides for a great variety of outdoor activity. Landscaping designed just to make the house impressive to neighbors and passersby is almost consistently an equally dull solution.

Formal outdoor areas should provide gathering places, a place to dine, a place to be alone and enjoy the presence of the full, rich plant life that abounds in a healthy garden. Fountains, sculpture, reflecting pools, paving, textures, benches, walls, trellises — all are elements that can transform an ordinary setting into a truly beautiful garden. Architecture is very important, but with excellent landscaping its importance diminishes. An ugly house with superb landscaping can still be satisfying, while a beautiful house with ugly landscaping is an eyesore.

The informal areas should provide space for games and play. How few yards have any special built-in items for these things to happen? Therefore, how few yards are used effectively? A swing set; a tree house; a slide; a basketball court; a lawn area for wrestling, football, and croquet; a playhouse; a wading pool; a sandpile are all elements that are irresistible to children. Many of these activities should take place in public areas, but they should also be provided in private yards, especially for the very young.

Another space, called an experimental creative area, is extremely significant. This area should be fenced off because it will undoubtedly achieve a high level of chaos. But it will also provide space for an immense amount of creative activity. A pile of bricks or concrete blocks, a bunch of old planks, an outdoor workbench, a sandy ground cover will provide the space for creative play. Building huts that can be dissembled and rebuilt for another need, digging lakes and canals in the

sand, building projects that are too big to do indoors — all are possibilities that this kind of outdoor area can provide. A noted British author of a book about playgrounds, Lady Allen of Hurtwood, maintains that in many of the larger housing projects and new towns, these areas are by far the most successful playgrounds.

Open Space

Public areas with playgrounds, football fields, barbecue pits, bicycle paths, tree-lined streams, and forest areas are amenities that are most vital in an urban setting. The trend of cities seems to be ever toward increasing densities; therefore the public open space will be the element that can ensure a neighborhood's remaining a healthy and vital place. Recently a sociologist maintained that children are not naturally drawn to the open green space but rather to the crowded sidewalk where the action is. One might falsely conclude then that green spaces are not necessary.

The real answer lies in the fact that children want to be in *active* areas, which simply reinforces the idea that the quality of outdoor space is determined by the number and variety of things to do in it. The children who left the green open space for the street only did so out of boredom. They did not leave an unneeded area; they simply left an area that failed. Often it is felt that a set of swings and a jungle gym create a playground. Single-function facilities will be used for a very short period because children easily become bored.

83

As with the interior, the exterior of a house should provide formal space, informal space, creative space . . . and pathways to open space.

And of course many standard playgrounds are so dismally ugly — a kind of plumber's nightmare — that even the kids are turned off.

Children who live in beautiful upper middle-class subdivisions with no open space have inherited an extremely drab, barren place in which to grow up. The general pattern for children in these areas is to have several friends whom they enjoy, and that is the extent of their close associations. If none of their close friends is available, they quickly resort to the television set rather than seek out other friends with whom they have slightly less in common. The great contribution of the open space is that it becomes a natural gathering place for a great variety of children.

(As a child I had the very good fortune to live next to a small but very adequate park. I remember that I did not waste time going to a friend's home to see if he wanted to play but rather I went immediately to the park, knowing full well that if he were free he would undoubtedly be there. As a result, I not only played with my good friends but also with children with whom I had definitely less in common, and often I played with kids I didn't particularly like at all, but they had the specific advantage of being terribly available. This, of course, is the place where possibly the most significant education occurs. Friends here are made, loyalties established, battles are begun and then settled, four-letter words learned. Some educators would have real-life situations occur in school, but this is the natural, spontaneous place for survival education to occur, and the schoolroom will always remain an over-structured, artificial substitute.)

Again, the basic philosophy is that design should be based on activities for people, and therefore landscaping should in its turn express this multitude of possibilities. To allow residential areas to be built without public open space in the immediate vicinity has been the great crime of land development in the past. Absurd as it may seem, it remains so today. Well-planned communities are still the exception. When is local government going to make public open space an absolute requirement rather than a friendly suggestion? When are we going to be able to enjoy the standard of public living that Europeans enjoy? In a zoo we go to great lengths to re-create natural settings so that the animals will relate well to their new environment. This is accepted as being important for the animals' physical as well as mental health. If only we would give our young and ourselves the same consideration.

It is indeed extraordinary that we as a nation lack determination to reduce overcrowding and create a healthy natural environment for the citizens in our larger cities. Our passivity is even more remarkable, considering the enormous number of people who express psychotic symptoms in the harsh, unnatural landscape of the overcrowded urban centers which are more and more exhibiting the symptoms of behavioral sinks.

We have inherited large-scale parks from past generations which have given us a false sense of solving the needs of the urban dweller. We should indeed be appreciative of these great landscaped areas, but as beautiful and spacious as they are, many times they are also miles away from the majority of the people. Significant open space must be part of the daily orbit of those who use it and therefore an integral part of the neighborhood.

Wonderful old cities in Europe are famous for their architecture, but the thing that makes them so successful for those who live in them is the way that nature has been used so effectively. There are parks everywhere, many subtly woven into the neighborhood fabric. There are outdoor cafes, there are duck ponds and flocks of ducks and pigeons to feed, there are playing fields for soccer and magnificent playgrounds. There are fountains, beautiful vistas, magnificent walkways, and there is an infinite variety of things to do, and thus city living is rich and rewarding. *Homo sapiens*, like other primates, just naturally responds to a lush natural setting. Certainly man in the twentieth century ought not be satisfied with less than a simple, honest, intelligent, inquisitive environment in which to live and raise his offspring.

CHAPTER SIX

Appliances, Materials, and Technology

"As long as design concerns itself with confecting trivial toys for adults, killing machines with gleaming tail fins, and 'sexed up' shrouds for typewriters, toasters, telephones and computers, it has lost all reason to exist."

Victor Papanek

Amerika has a great freeway system because of the combined influence of the automobile and oil industries. Many critics of the war in Vietnam maintained that we were involved in a controversial war in order to satisfy the demands of the military-industrial complex. Certainly the two previous statements grossly oversimplify a much more complex situation, yet they point out a definite characteristic of a highly industrialized society: that an extremely wealthy industrial complex has an undue influence on the direction of government. It also exerts its influence on consumer demands by clever advertising, but seldom with any deep concern for the social merits of its products.

Appliances

For the reasons stated above, we find ourselves filling our homes with a lot of technological gimmickry which provides a minimum of human comfort at considerable cost simply to appease an aggressive appliance industry. This is not to imply that all appliances are of little value, but rather that their value could be and should be measured in terms of what they actually accomplish. A good example is the enormous difference in cost between a normal drop-in range with a self-cleaning oven (which is an expensive but reasonable extra) and a super-special model with little knobs and flashing lights and an extra warming oven, with all kinds of frivolous things going on. Why go the extra cost? The difference could mean something of true personal value for the family.

Another unique but rather expensive item is the built-in vacuum system. Here is a product with a dreary history of unkept promises, yet with excellent advertising it has managed to stay reasonably alive. It seems ridiculous to install the heavy equipment necessary and long runs of duct work where a small machine on the end of an extension cord (with a little healthy exercise required on the part of the housewife) is so much more efficient. A significant function of technology is to relieve man of his dreary, repetitive work to free him to do more significant work. The built-in vacuum system at great expense has really accomplished very little. (A curious phenomenon among clients who are bad housekeepers is that they feel that appliances and other mechanical devices are going

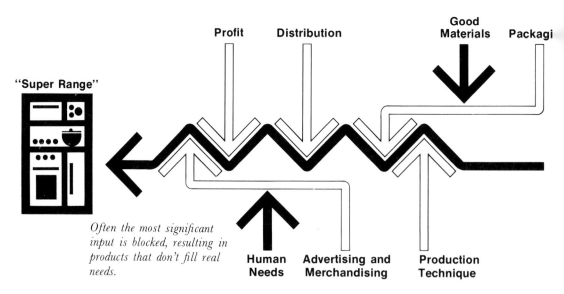

Often the most significant input is blocked, resulting in products that don't fill real needs.

to "save" them and somehow miraculously turn them into meticulous housekeepers, but this optimism is not warranted, for the miracle seldom occurs.)

The low-voltage, touch-plate electrical wiring system is another grossly overrated item. The merit of having a panel of little red lights that give a warning if any lights are on anywhere in the house is a somewhat dubious luxury. Low-voltage wiring is of course much more expensive and is therefore immediately attractive to the consumer who finds fulfillment in pressing a button rather than undertaking the exhausting task of flipping a switch. He has the added advantage of having to pay for high maintenance bills because of the nature of this highly sensitive system — or at least that is what electricians say. There are a few valid applications for this product, but it is a technological development that has been manifestly misused.

America is also the most overplumbed nation in the world. Home builders with little imagination have used superabundant plumbing as their way of creating luxury in a house. Most Americans enjoy at least as many baths as they do automobiles. One is almost forced to have concern for the basic physiology of people who need facilities at such close intervals. Somewhere along the line it was decided that parents and children ideally should not use the same bath, and of course this concept was given some encouragement by plumbing manufacturers. What was at one time a luxury now has become a minimum requirement. Upper middle-class Europeans and Japanese are satisfied with one car and one bath, and rather than being mortgaged to the hilt to pay off double facilities, they have more money to travel and engage in other mind-stretching activities.

Galbraith, the renowned economist, maintains that the basic patterns of consumption of goods are not based on need but rather on the productive capacity of industry which then must unleash a fearsome advertising program to market that which has been produced. On the other side of the ledger is the imminent fuel shortage which threatens us. Galbraith points out that even at times when demand is low and supply is high, prices are not lowered substantially, but rather a new sales campaign is begun to encourage consumers to absorb the oversupply; and of course built-in obsolescence guarantees this same manufacturer a vast market for replacing appliances, more evidence that the consumer is not given the highest consideration.

Indeed there is nothing wrong with extra plumbing, built-in vacuum systems, low-voltage wiring, and so forth, if the homeowner can afford them and can also afford all the basic amenities that will give him a good home. The unfortunate thing is that so often these extras are obtained at the expense of the things that matter most.

Generally, the worst designed houses, those that function poorly for raising a family, have the greatest amounts of technological equipment. Two baths are not a must in a home. A range with knobs and flashing lights is not a must. The creative spaces and vital gathering areas which really determine the quality of living should establish the norms of planning rather than the productive capacity of industries openly seeking new markets.

A typical 1,500-square-foot house will cost anywhere from $1,200 to $1,800 to air condition, but cooling often can be solved for considerably less in much of the western part of the United States with evaporative coolers or water coolers. But there are much better and more natural solutions. It costs only about $100 to $150 to move a very large tree. With a

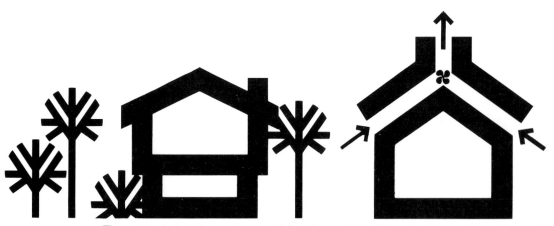

Trees not only help keep a house cool but also reconvert polluted air.

A double roof with an exhaust fan provides cooling with minimum power use.

budget of $1,200 to $1,800, a number of large trees can be brought in and placed strategically around the yard, and the cooling problem can be solved with the added benefit of magnificent foliage.

Another possibility is building a double roof with an area underneath the top roof for shaded ventilation. There are many more *natural* answers to cooling that have already been discovered and many that will be; but in a technological society we tend always to look for machines to solve our problems. And of course in a technological society to look elsewhere is a considerable heresy. Natural solutions would almost consistently be less expensive, and this again allows more budget for the things that really count. But even more important, natural methods are part of the ecocycle, helping to preserve the balance of nature.

The deterioration of architectural design can be attributed to many factors, but a very significant one is the immense cost of the mechanical equipment going into houses. When this cost reaches somewhere between 30 and 40 percent of the total budget, it can readily be seen that there is little choice left but to solve the problems of design in the most direct and least expensive way. If we had had air conditioning in the thirteenth century, we would never have had Gothic cathedrals; 30 percent of the budget would have been lost in mechanical equipment. We would have lowered the ceilings

95

Windows strategically located can receive solar heat during the cold winter months, thus conserving fuel.

to reduce the volume and then would have suspended the ceiling to hide the duct work. This example is pretty far-fetched, but it certainly points out the problem. Certainly mechanical conveniences have a contribution to make, but so do architectural elements that we have discarded as too expensive.

We must insist that mechanical devices and appliances are justified only in terms of what they accomplish for human benefit. Properly applied they should free us, not further enslave us. We are fortunate to have them, and we should use them with discretion and good common sense.

Perhaps our greatest hope lies in the young designers who are committed to producing elements at the lowest possible cost in order to reach the greatest number of people. Many are committed to a new direction called bionics, which is simply studying nature and resolving technological problems in a natural way. Excessive power demands, using the old technology, will only leave us immersed in our own pollution. Bionics hopefully will create a technology that will be naturally integrated into the ecocycle.

Sewage disposal has always been a big problem, one that inhibits home building in many areas of the country which lack public systems. Unquestionably the ideal solution lies in an individual home treatment plant where the decomposing matter can eventually break down into a harmless

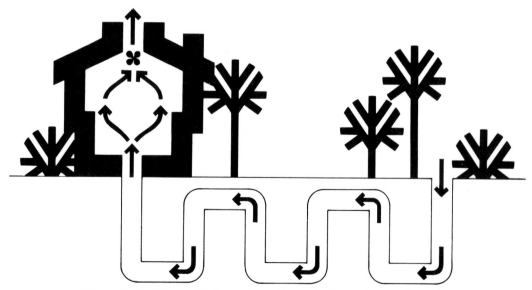

The earth itself is an effective coolant. Air can be drawn from a shady spot, ducted through 65-degree earth, and exhausted through the house.

powder, while the fluids can be recycled into the water source. An added advantage is the additional electric power created in the release of energy realized through the process of decomposition. This type of system is already in existence, and although it is in need of many refinements, it has great potential.

Our sources of power in many parts of the country have become undependable. An individual generator powered by a comparatively pollution-free motor using kerosene for fuel is already in existence and is used by many people in outlying areas. Transporting power over great distances causes a tremendous power loss and is thus an inefficient way of solving power needs for many situations. The individual generator is a promising power source for the future.

Both of these systems are technological possibilities that are anti-status quo, yet both, if properly handled, can more adequately appease the demands of ecological balance — and they are good examples of mechanical devices based on a common-sense evaluation of man's genuine needs.

Materials

Many people comment that it must be terribly exciting to be an architect today because we have so many new materials to work with. We do have a fantastic array of new materials, but the great majority of them are pretty dreary. Today architects yearn to use materials that are hundreds and (some) thousands of years old; but presently in a technological society, their use has been greatly curtailed because of cost. A brick floor, heavy hardwood paneling, a stone fireplace cannot be surpassed for beauty.

97

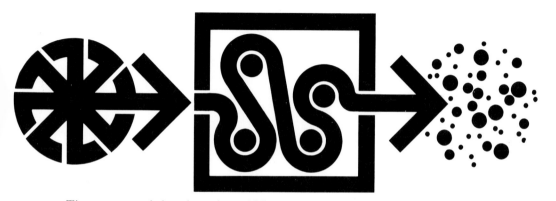

There are many design alternatives which would reduce our dependence on machines which burn precious fuel and create destructive pollution.

Sometimes we find a new material that is rather good, but the manufacturer's recommended applications are unbelievable. Floor coverings are a good case in point. Vinyl tile is a good minimum-cost material that works well where a hard surface is advisable. But tile manufacturers in their national advertising show vinyl tile in extraordinarily bad applications. For example, they show tile in a beautiful living room that looks very lush except for the tile floor, or they show a master bedroom that has a warm and inviting character except again for the chilling effect of the tile floor. Why not show tile in a kitchen or a bath or a playroom where it really belongs?

Ceramic tile ads are equally undiscriminating. They display ceramic tile on living room floors, which in itself is questionable, but they do not stop here. They go on to show tile on the walls, tile fireplaces, a built-in tile couch, a tile tabletop — tile all over the place! The viewer is forced to turn away from an overwhelming visual overdose.

It is unfortunate that a product cannot be shown with applications that are both logical and appropriate. Carpets once were reserved for living rooms, formal dining rooms, and bedrooms, but manufacturers cannot be satisfied with 75 percent of the floor area of the house. Now we are told that carpet is the best possible floor covering for kitchens and bathrooms. Regardless of how easy a carpet is to clean, this is a ridiculous application for the kitchen. Minute particles of food spilled on a carpet are going to remain. A buildup of these particles over the years cannot be considered the ultimate in hygiene. The superb misapplication, of course, is the bathroom, where little kids are expected to hit a 100 percent average! The housewife who detests cleaning around fixtures is eliminating that chore, but she must be responsible for a carpet that will achieve a fantastic bacteria count.

The most heralded of the new materials are the plastics. We use laminated plastic counter tops, and some plastic glues and finishes, fiberglass bathtubs, and plastic imitation marble counter tops. But we are still building essentially wood and masonry houses. The plastic house on display for so many years at Disneyland was well designed, but it was never produced for the marketplace. The white, sterile finishes were perhaps the reason — for with all its quality it lacked the feeling of warmth that many people find necessary for a home. Wood, brick, stone, and tile still provide this kind of magic. Plastic at its best is still a rather slick and barren material. When it is made to look like another material, it is abominable.

Prefinished hardwood paneling is neither less expensive nor equal in quality to unfinished paneling. Industrial finishes can hide many imperfections, but too much is lost in the process. The skilled carpenter or painter can bring out the depth and quality of the grain that the industrial product is forced to hide. Aluminum siding, the almost universal answer for maintenance-free surfaces, is also grossly overrated. Even if it did not have the problem of denting, its sterility as a material alone should disqualify its extensive use.

Industry has developed a core of product designers who are not involved in designing a finished product. This is why we are in the strange position of having an infinite variety of truly ugly materials to choose from. Yes, there are new

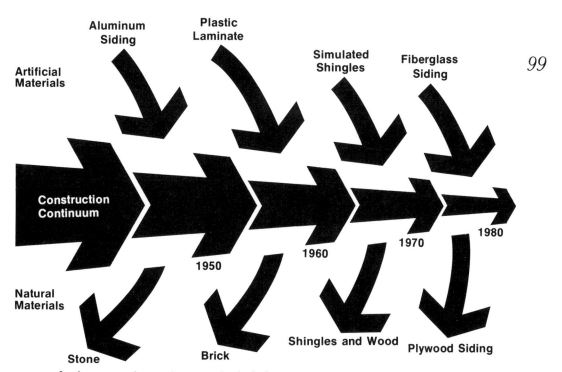

In the construction continuum, technological progress means rejection of natural materials in favor of extruded products. The result is not freedom but restriction.

materials, but they have advantages and disadvantages just as the old materials we have had for years. That is, with one exception — the old materials have a natural richness that has never been equaled by the new.

Housing Technology

The housing industry has received considerable criticism aimed at its monumental inefficiency. The idea that we are building houses the same way today as we did in the past happens not to be true. Many changes have occurred. For example, the use of plywood rather than sheathing and gypsum board rather than plaster have precipitated revolutionary changes. The normal house today is much better built than the house of the past, contrary to what most people think about the sturdily built homes of yesterday. The only superior element in the old houses is the interior finish. Structurally they are almost always inferior, even the very large and expensive ones.

Heavy industry is now interested in the housing market. It will undoubtedly come up with efficient solutions, but its solutions will have very definite limitations. The one thing that must be said for the standard way of building is that the possibilities are infinite in terms of exterior and interior design, as well as choice of materials, height of room, structural system, and so forth.

Today there are many houses that are built at the factory and then trucked to the site. The problem of transportation is, however, the most limiting of all factors, for the section of the house is restricted to a twelve-foot width, and the pitch on the roof is also narrowly restricted. The height of the total unit cannot exceed the height of the average underpass to be encountered while delivering the house to the site.

Many exotic systems are sure to emerge, and undoubtedly a few will work well at a good cost; however, considerable building is going to be carried on in the same way we are building today. Conventional methods will primarily be used for more expensive housing because they still have a reasonably good cost for custom building. For low-cost housing we are going to have to take advantage of some form of prefabrication or industrialization in order to gain the efficiency necessary to bring costs in line with what people can afford.

The forms are being decided now. One solution is the factory-built home, which can be either a mobile home or a home built in several sections then moved to the site and placed on a permanent foundation. Production techniques and costs are similar for the two, but the final product is somewhat different. The mobile home turns out to be a bad long-term investment, while the permanent housing is quite good. Thus it makes little sense to resort to cheap, temporary solutions.

A common misconception concerning the production of factory-built housing is that it is a highly industrialized process. Actually, the main efficiency is in teaching unskilled laborers to do simple production-line jobs — and of course paying them unskilled-labor wages.

There is as yet little fabrication or production by machines. Mobile homes are actually being built as they have been in the past, and there have been few revolutionary fabricating techniques. Machine fabrication of component parts nevertheless still has some exciting possibilities. Only mass production by machines can really bring costs down. Whether the assembly of the mass-produced components occurs in the factory or at the job site is not too important, except that job-site erection allows for a wider variety of design possibilities. We must use mass production, we must use materials more efficiently, and we must encourage skilled labor to triple its capacity. Where six carpenters can now produce fifteen houses a year, they will have to produce forty-five or more in the future. If we can develop systems that trained craftsmen can erect in only a few days and finish in another two or three days, we can achieve fantastic levels of production at reduced costs.

Europe is now in the third generation of prefabricating, while the United States is in its first. Europe began by using the modular approach (similar to our large factory-built sections), but has slowly abandoned the modular for mass-produced wall and floor sections that are assembled at the site. European nations have carried on an extensive research program in trying to determine sound prefabricating systems.

Therefore, when they conclude that the modular is not the best approach, and when the Soviet Union (the last to discard the modular) has also finally conceded, Europe stands unanimously in favor of prefab components erected at the site.

The United States, with very little subsidized research until recently — and an utterly disorganized and costly effort at that — is almost universally involved in the modular approach discarded by Europe. Perhaps our way of building is different. Perhaps we can make the modular work. But so far the facts are not in its favor. It would indeed be tragic if it were to take us three generations to arrive at the same conclusion as the Europeans. We have a fierce nationalistic pride in our technology, but here is an area in which we must not only bow to Europe, we must also thoroughly examine what they have accomplished. If the necessary research has already been done, we must take advantage of it.

In all of our technological experimentation we must not be mesmerized by cheap solutions or exotic structural techniques and forget that the main function of low-cost housing — and for that matter, all housing — is to provide an inexpensive envelope for exciting and creative space. To be satisfied with clever technical solutions alone is certainly to lose sight of the things that matter most, and human storage is again the inevitable result. Technological breakthroughs will occur but it is hoped they will remain subservient to the sociological breakthroughs which will also occur and which will undoubtedly be more significant. What we need is inner-directed design by designers committed to solving problems in the most direct, logical, and natural way, but with a deep concern for the quality of living that good design can help to provide. This will certainly be a departure from the other-directed design that is primarily subservient to unconscious external pressures exerted by manufacturers and suppliers who are devoid of any real concern for solving significant human needs.

Technology ... Is It Merely Apocryphal?

Technology, after all is said and done, remains a woefully limited blessing. Though it has freed us from drudgery, it simultaneously has displaced the craftsmen. We have been overwhelmed with a vast quantity of materials, but we have been forced to search desperately for quality. It has given us the power to subdue nature, but too often through misuse, we have bludgeoned nature to death. As a result, we have more free time, but we have created a barren environment in which to enjoy it.

Technology has never been, nor will it ever be able to do the important work of society. Intelligent people must constantly be involved in a reevaluation of their value system and be equally dilligent in transferring the value system to the next generation.

It takes a great amount of work, effort, and people to create a healthy environment, and that time won from our mechanical slaves must be used to create a balanced environmental structure, as well as the physical elements that technology has made obsolete. A stone wall, a brick pathway, a cobblestone drive, a complex wood lattice, a beautifully carved door — all make their contributions to a humane environment.

If technology has destroyed the crafts, let the community restore them. Economic obsolescence does not mean social obsolescence. We can still decide what is good for us and our children and then go about doing many things ourselves. Because we cannot afford a stone mason does not mean that we cannot have a stone wall. We can build it ourselves. Let us not allow technology to drain the life out of living. Technology's limitations are *its* limitations not man's, and only man — subdued by technology — will accept these limitations. The great variety of crafts that were economical requirements of the past can now serve as social requirements of the future.

CHAPTER SEVEN

The Housing Industry

"The central function of architecture is thus to lighten the very stress of life. Its purpose is to maximize man's capacities by permitting him to focus his limited energies upon those tasks and activities which are the essence of human experience."

James Marston Fitch

I‍t is a fortunate period in history when the various forces at work so combine that great achievements are realized. During the Middle Ages, the tremendous influence of the church and the subordination of the individual to the will of the church channeled all the creative thrust of the age into the great religious architecture of the Gothic period. In later centuries, during the Renaissance, the individual began to emerge once again, and Western civilization experienced a great period of discovery in a multitude of areas, including architecture.

When all the elements seem to fall properly into place, great things are realized — but this is distinctly *not* the case in recent history of housing in America. Here freedom of the individual in many ways has probably never been greater — but certainly we have not experienced a great period of residential architecture. On the contrary, we have built "corporate cathedrals" and "garden factories," but we have experienced a distinct decline in residential architecture. It follows logically that we turn to the housing industry for some explanation of our dilemma. Certainly there are many socioeconomic pressures that can be held partially responsible, but the blame finally must rest squarely on the shoulders of those who produce our housing.

The housing industry is comprised of many separate groups. Among them are land developers, builders, financial institutions, the Federal Housing Authority (FHA), labor unions, architects, and local government agencies. Together these groups have produced a tremendous number of homes, and they have benefited greatly in the process. Unfortunately, the homeowner has not shared equal benefits. Perhaps the primary problem is that the many specialists who comprise the industry have had little knowledge of the services the others perform. The builder has had little concern for the things the architect is trying to accomplish and has seldom hired him. The architect is ignorant of the requirements of the financial institutions and the builder. Individually, each of these is capable of doing a good job in his own specialty, but the integrated process of producing a well-built, well-designed house at prices that many can afford has rarely been achieved.

In order that we may more thoroughly comprehend the composition of the housing industry and its problems, it is necessary to analyze the activities and services of these specialists.

Land Developers

Land is purchased, subdivided, and sold for a profit by the land developer — a process that has resulted in some of the most spectacular inflation that has occurred in recent history. Building costs for many years remained fairly constant, while land costs rose steadily. Since 1956 there has been a 400 percent increase in land costs, while construction costs have risen 100 percent. Another source of trouble is that land is developed in small pieces of from twenty to fifty acres, but never in large enough pieces to plan and develop a total community.

Land developers first master-plan an area, then provide the capital to improve it. They are responsible for placing the curb and gutter, which add approximately $2,500 to a raw land cost of let us say $1,000 per lot. When the same lots are sold for from $9,000 to $12,000 each, the extremely attractive margin of profit can be seen. There is no *actual* value in subdivided land. What really gives value is the building that is placed upon it, but often land costs are so high that builders are forced into substandard building to make a project work economically.

In his search for inexpensive ground, the developer is generally forced to purchase land in the wrong places, too far from the city, where there is no community. This of course creates a problem which is compounded for the taxpayer who must bear the financial burden of delivering public services to these outlying areas. Generally speaking, land developers have been the most shortsighted in the home-building process. It is unfortunate that those who have contributed the least have often made the greatest profits.

Financial Institutions

The financial complex, comprised of insurance companies, savings and loan associations, and other banking institutions, arranges mortgages for homeowners. People trained in finance have generally had little concern for design. They have also had virtually no concern for community amenities such as parks and recreation facilities, and have seldom been willing to give a building a higher value for having these elements available as part of the immediate community.

This esthetic and social illiteracy has perpetrated the financing of an unfinished product; landscaping and built-in furniture are elements which they have normally been unwilling to include in the mortgage of a home, although some frequently include carpets and drapes in these long-term contracts. Yet all these features are absolutely necessary to make a house habitable. One might compare this practice to the automobile manufacturer marketing a car that was not complete enough to drive.

Another problem is that financiers have a profound fear of innovation, feeling more secure with proved concepts.

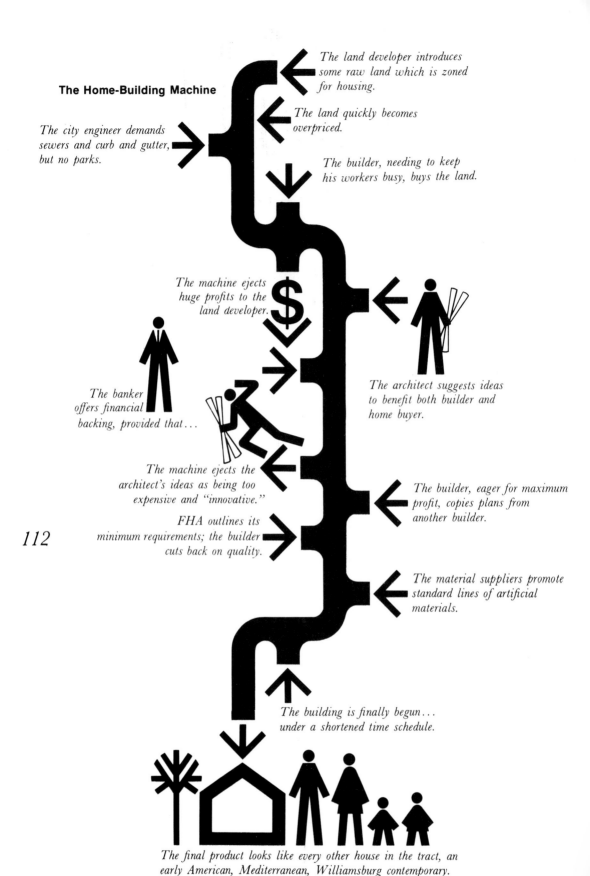

The Home-Building Machine

The land developer introduces some raw land which is zoned for housing.

The land quickly becomes overpriced.

The city engineer demands sewers and curb and gutter, but no parks.

The builder, needing to keep his workers busy, buys the land.

The machine ejects huge profits to the land developer.

$

The banker offers financial backing, provided that...

The architect suggests ideas to benefit both builder and home buyer.

The machine ejects the architect's ideas as being too expensive and "innovative."

The builder, eager for maximum profit, copies plans from another builder.

FHA outlines its minimum requirements; the builder cuts back on quality.

112

The material suppliers promote standard lines of artificial materials.

The building is finally begun... under a shortened time schedule.

The final product looks like every other house in the tract, an early American, Mediterranean, Williamsburg contemporary.

The men who provide the money have often been a stumbling block in the past because of their extreme caution. As a result, progress in housing has been very slow.

The Federal Housing Authority

FHA is the government agency responsible for insuring mortgages provided by financial institutions. The theory behind the process is a good one. It works much in this manner: Bankers want from 20 to 30 percent for the down payment on a new house. FHA has been willing to insure loans where all that is required by the owner is a 10 percent down payment. This insurance therefore covers the difference between what the bank is willing to loan and what the buyer is able to furnish as a down payment. There is a great variety of programs, but all are essentially variations of this.

FHA, begun in the early 1930s, established some strict minimum standards regarding structure, materials, and other construction features. The problem emerging here is that when standards are established, one can be sure that those involved will do the absolute minimum to meet them, and this is exactly what has happened. Ironically FHA, by enforcing its minimum requirements, has been responsible for some truly minimum housing. In many places the quality of FHA-insured homes has not been nearly equal to those completed under conventional loans. Good architectural planning and design have never been strict requirements, and an extraordinary deterioration of design has occurred under the sponsorship of the broker-builder and FHA.

Yet another problem is the extreme amount of red tape that has forced the more creative builders to shy away from FHA channels. As an example, it is interesting to note that the Department of Housing and Urban Development has had to establish Operation Turnkey to circumvent its own bureaucracy. Operation Turnkey makes it possible to simply go out and get a project built, then turn it over to FHA — thereby bypassing many bureaucratic requirements.

Perhaps the greatest weakness of FHA has been the lack of community requirements in its building regulations. Not until recently has it realized the value of what happens *outside* a house. It has never determined community needs such as the number of houses that necessitate a playground or how many more indicate the need for a playing field or swimming pool. It has consistently placed such things as curb and gutter requirements before bicycle paths. The tragedy is that today FHA is conscious of its past errors yet is not only allowing but *subsidizing* many projects that are being built in exactly the same way.

Another serious difficulty for the home buyer is forced conformity brought on really through FHA red tape. It takes so long to get a home model accepted that generally a builder will establish five basic models and build two hundred of them. Certainly we cannot feel that two hundred home buyers are having their needs met by five basic designs, yet the home builder is forced into this pattern.

Another questionable activity of FHA is its involvement with expensive high-rise apartments at the same time it has not been meeting the needs of the poor. A lot of noise has

The widely divergent goals of government regulators, land developer, builder, banker, city planner, and craftsmen make good housing hard to achieve.

been made about housing for the poor, and many programs have been established, but there have been precious few low-cost housing units actually built. FHA has finally realized the desperate need, and today a serious effort is being made to meet that need — but this key agency is years behind, and each year finds it farther behind.

Through its Operation Breakthrough, FHA looked to heavy industry to solve the problems of mass housing. The frightening thing is that this would undoubtedly have produced an extremely limited variety of housing; also, if heavy industry were to tool up for mass production, it must be assured of continual demand. We have a certain building capacity now, and we must double it. This can be done by prefabricating, by streamlining building codes, by utilizing continual technological advances, and by simply making the existing industry more efficient.

Although it may sound like a contradiction in terms, FHA has had extremely creative bureaucrats. An immense

A good house in a good neighborhood must become a common goal; then regulators, developers, builders, bankers, planners, and craftsmen must learn to cooperate and function as a team.

amount of research and lot of writing have been done, but there remains a vast difference between what has been proposed and what has been achieved. We must be able to depend upon a creative bureaucracy to take a broad view of the problems in the housing industry because it is at the bureaucratic level that high-sounding policy and legal requirements must be carried out.

Builders

There are men who are given the power of kings — the builders. They determine what is to be built, how and where — and for whom. Builders are the biggest promotional force in housing today. Generally they are brokers and organizers and not highly trained craftsmen. Too many are unimaginative, poorly trained, gimmick-oriented copy artists. Each generation leaves its mark in the kinds of buildings it provides for coming generations. It is unfortunate that the housing industry has not attracted the quality of individual that this vitally significant effort deserves. Builders are a product of FHA requirements, a response to its bureaucratic system. They tend to give the very least, yet aggressively pursue the greatest profit. They have forced architecture out of housing design and have created a square box with holes punched for windows!

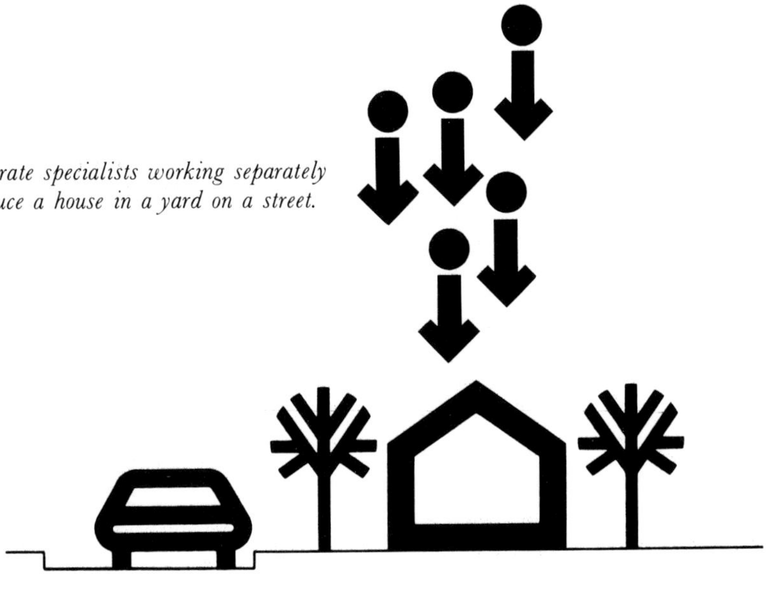

116

Separate specialists working separately produce a house in a yard on a street.

Certainly there are some exceptional builders who are socially concerned and who have tried to give their best. The National Home Builders Association has been an involved group whose leadership and direction have too often fallen on deaf ears. But generally speaking, there has never been a less imaginative group of men — as witnessed by street after street of their dreary products.

Labor Unions

Labor unions too have become an irresponsible element in the construction industry. They have virtually lost their social concern, and labor has become a narrow, self-centered, reactionary institution. It would seem evident that where there is such an outstanding need and such an enormous amount of work to be done, the workers should not want to limit production in any way by insisting on outmoded requirements and procedures. The unions no longer seem to have a particular concern for the quality of the work produced by the various crafts. The restricting of minorities from joining their numbers is not only unforgivable in terms of civil rights, but has severely curtailed the labor supply when there is an insatiable demand.

117

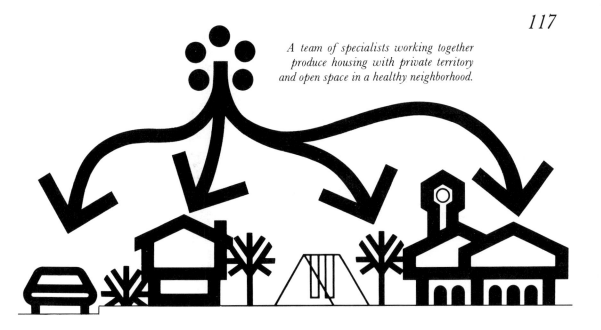

A team of specialists working together produce housing with private territory and open space in a healthy neighborhood.

Architects

In the past, architects have found a considerable amount of their work in housing design. This was true until just before World War II, and most prewar housing developments achieved a reasonably high level of design. This is obviously not the case today. The tragedy of the architectural profession has been its lack of concern for how the nation is housed, and its fixation with a 6 percent fee; the latter, of course, generally prices architects out of housing design.

Architecture schools have turned out specialists who go through long and complicated design procedures to arrive at a final solution. This has proved to be a successful technique in handling large commercial and institutional buildings, but there is neither time nor budget for this process in housing design. Where the architect has been unwilling or unable to aid the builder or less affluent client, the house "designer" has attempted to fill the gap. With some notable exceptions, this group has been a barren source of architectural design. In fact, the combination of an unimaginative builder and an equally unimaginative designer has produced unpropitious results.

A great burden of guilt falls on the shoulders of the architectural profession, which knew full well what was happening but abdicated its responsibility.

Local Government

Our local agencies have felt that their highest calling was to police the builder through building codes and zoning ordinances. Even with these tight controls, some terrible things have been allowed to happen simply because there were no rules against them. For example, local building codes have specific requirements for homes but virtually no requirements for mobile homes. As a result, their ofttimes excessive requirements have increased the cost of standard housing, while allowing cheap, substandard, often hazardous housing in the form of mobile homes to be provided free of inspection requirements. At the same time, they have placed a heavy tax on the homeowner, while the mobile home has remained relatively tax free.

In this indirect way they have encouraged the mobile home and discouraged better housing. Government has felt that its duty was fulfilled when it insisted on curb and gutter and sidewalks for newly developed areas, while it disregarded planning requirements for parks and open space as well as neighborhood centers. Fear of rising taxes has of course been the reason for not insisting on these elements.

If it were not for these agencies, however, we would not have the parks and libraries we do have, and socialistic as it may sound — and expensive as it will certainly be — the further responsibility for insisting on the kinds of elements that comprise a healthy environment falls on the shoulders of local government. We are taxed by local government to provide needed services. The planning department of most cities is inadequately staffed and is void of the kind of power it should have if it is truly to influence the community planning. As a result, we are not receiving these crucial services.

This has been a brief but critical analysis to delineate and accentuate the problems. There are many specialists in each area of responsibility who do not fit the negative descriptions, but they have not been the majority...nor have they been in the positions of power, because the landscape is filled with grim little boxes that remain the damning evidence.

Positive Coordination Could Pay Off

Now let us investigate how the same forces can work more effectively together. First, local government must become a creative force. Its agencies must establish some basic planning requirements. For example, each block should have a playground and a series of blocks should have a playing field, and each neighborhood should provide the recreational and social possibilities necessary to a healthy environment. Some recreational and adult education facilities can be tied in with the public schools; this practice is known as the community school concept and is supported by many valid arguments. Perhaps a specific improvement district can be established to pay for playgrounds and parks. (The typical community will support a bond issue for a sewer system or a highway system and yet totally disregard some basic and much more important direct human needs.)

Government can also do a great service to a community by keeping land costs to a minimum; instead of taxing *improvements* on land, it should tax the land itself on the basis of its highest potential use. This would force land developers to move fast or sell their land quickly at more moderate costs in order to avoid a high tax on undeveloped land. Taxation can be used as an effective planning tool. Instead of allowing builders to build just anywhere, local government should insist on master-planning new developments into completed neighborhoods as well as planning for new towns, and then become the catalyst in the creation of these new towns — in the manner of the London County Council, for example, which has been responsible for creating entire new towns of fifty thousand and a hundred thousand population and then connecting them to the parent city with a rapid transit system.

Despite attempts of large corporations to establish new towns, this should really be the function of local county government because good planning requires continuity. The

120

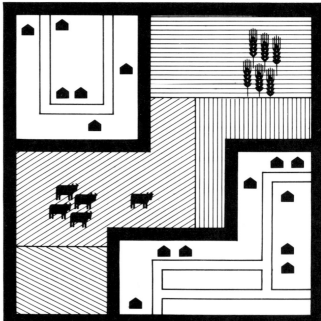

Disorganized development allows builders to create small self-contained housing areas that have no open space and no relation to one another.

county could do a more effective job because land profits would not be the great motivating factor. A creative bureaucracy composed of architects, planners, sociologists, and psychologists could give the creation of communities the broad view that such a socially significant effort demands and indeed deserves.

The builder is going to have to fulfill his social responsibilities. His great power must be used to provide houses that are well planned and well designed and to place them in communities that provide the recreational, social, and educational amenities which ensure a healthy, self-sufficient community. We must somehow rid ourselves of the irresponsible promoter type of builder and put our support behind men who understand the term "environment." The new builder should be a team of specialists in the planning, designing, financing, and building of completed neighborhoods and communities. This type of building organization is now emerging in many parts of the country. Local government, FHA, and banks must insist on this type of complete capability.

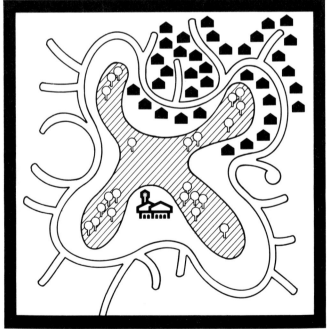

Cooperative development creates a master plan, road patterns, open space, and village centers so that healthy neighborhoods result.

The architect must become involved again in housing design because it is by far his most significant opportunity to influence the quality of living, and we are in desperate need of new ideas. He is going to have to develop new procedures whereby he can provide a service to those who have been unable to afford him in the past. He must reach out to other disciplines such as sociology and psychology to give his designs a vital direction. A society that has little pride in the way it expresses itself through art, music, drama, and architecture has little concern about its own survival, for it has little worth preserving.

Labor unions are going to have to eat some of their sacred cows. We need production, we need it desperately, and we need it now. The argument for limiting production to give security to workers is ridiculous in a period of enormous population pressure.

The Federal Housing Authority, like local government, is going to have to become a creative force. Its *minimum* planning requirements must include such human necessities as green space and community recreational facilities. FHA must reach out and appeal to the creative builder and the talented architect, and it must work closely with local government to become a strong promotional force. It must take its responsibility seriously and not be willing to give in to the same old way of doing things, or become enamored of quick and easy solutions.

FHA sets the pace for the entire industry and has become its conscience, but a dynamic philosophy of what housing should accomplish is still forthcoming. After all, bureaucracies in Europe have achieved some great things in housing. Why can't it happen here?

Financial institutions must also become educated in the best thinking concerning community planning and seriously back those builders who are committed to good land planning and architectural design. They must work hard to create the kinds of total financing which will take into consideration landscaping and community needs, for only this will allow the builder to market a really finished product. They must be willing to think anew and not fear the revolutionary process that we must go through if we are to succeed in working out the basic economics of a heathly environment. If we are to meet our housing needs, the institutions responsible are going to have to make some radical changes, and new specialists are going to emerge.

Possibly the real hope lies in new corporate organizations to handle the entire process of land acquisition, planning, and development, and the architectural design, financing, construction, marketing, and management — thereby circumventing the irrational processes that have become so entrenched in the status quo.

Certainly the immense social significance of housing at least equals but undoubtedly surpasses that of education. Think of the great variety of trained staff required to run a large education system. Housing needs this same kind of total effort.

Organizations that can provide this kind of *total* service should be given the greatest support by government and finance, for their chances of succeeding are far greater than those of a splintered, limited service at the mercy of a multitude of irrational forces.

It is indeed rare when these varied forces so combine that great things are achieved, yet in an era as unstable and revolutionary — and as promising — as ours, we dare not settle for less. We must begin now.

CHAPTER EIGHT

The Neighborhood

"The strangers-in-our-midst phenomenon tends to clamp down on the tribal-sharing, social mixing atmosphere typical of smaller communities. Defensively, the families turn in on themselves, boxing themselves off from one another in neat rows of terraced or semi-detached cages."

Desmond Morris

I n the vast urban complexes that we continue to produce and that are beyond our capacity to comprehend, is our only security to be found in our own limited, private world and the small remote community of our choice, such as our church or country club or auxiliary? If we feel satisfied that this tight little world solves the needs of our children as well as ourselves, we are indeed deluded, for an element of immense significance is lacking. A community limited in scale and number, bound together by a powerful group commitment, is that missing element.

In Chapter 7 we discussed the housing industry and its lack of any consensus of direction as to what housing should provide. The industry has produced a dreary product architecturally; more importantly, it does not fulfill significant human needs. What is true of housing is unfortunately also true of community planning. There is considerable agreement about how to design street intersections, how to separate

129

neighborhoods from industrial areas, and how to provide a good circulation pattern. But there is little agreement about what a good neighborhood should provide, what size it should be, how many people it should have, how it should relate to the larger community, and how it should be governed. The planning profession, unlike the housing industry, is not deserving of an indictment simply because it is a new field and needs time to formulate a direction. Planners are searching for solutions to these problems. But we are discussing here the sociological and political problems that have thus far not been part of the planning discipline.

Planners have created some new towns recently both in Europe and in America, and they have tried to create well-balanced communities. Their success can only be adequately assessed at some time in the future. Already, however, criticism of these new towns is forthcoming as they begin to exhibit some of the same kinds of problems that existed in the older cities. The planners of the new towns can be found

In apartments, individuals who share common entrance
alcoves tend not to socialize. This is an excessive propinquity.

guilty of extreme optimism in their ability to provide an ideal community. But if they have not achieved the ideal, certainly these new communities are almost universally far superior to the unplanned, unconscious sprawl they were designed to replace.

Yet there is still something lacking. What we are talking about is the ideal community, and it becomes apparent that we are not attacking the problem at the proper level. The individual cannot fully comprehend a new town of a hundred thousand people any more than he can an old town of the same size. Earlier we talked of scale and orbit and the genetic inheritance that renders us simply incapable of relating to large, anonymous groups of people. We have, through our slow evolutionary process, been programmed to relate to and to cooperate with the small hunting group. As a result our small communities of two thousand to five thousand people seem to function reasonably well internally, and seem to somehow respond naturally to the needs of the inhabitants.

Small communities can have serious problems, and it is senseless to romanticize them into something that they are not. At the same time, they have the great advantage of functioning on a comprehensible scale. And again we must emphasize that the small community more closely resembles the group life for which *Homo sapiens* was programmed. The small community or neighborhood becomes the critical element. The new town, like the old, will only be as healthy as its neighborhoods.

Healthy neighborhoods can exist in Harlem and Los Angeles as well as in Pocatello, Idaho, and Columbia, Mary-

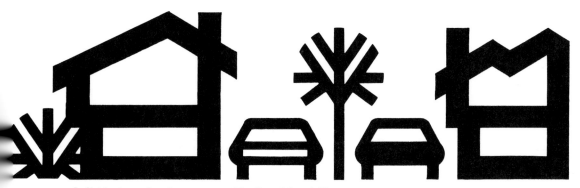

Individuals tend to become most friendly with neighbors whose driveways adjoin theirs. This is a healthy propinquity.

land. The neighborhood is in scale with man, because here he satisfies his primary biological and psychological needs. Here he has his territory where he and his mate will provide security and education for their offspring. Here, within a cooperative territory, he unites with his neighbors to provide a more secure place for his progeny. Here in the commonly shared territory, the young learn the advantages and benefits of cooperation. Here they achieve identity, security, and stimulation. Here, exposed to a great variety of practical, everyday experiences, the young receive their most important education — that which relates to and ultimately will determine the survival of the species.

The neighborhood, then, is immensely significant. It is in scale with man. It is the civilizing force. Without it we remain immersed in the public chaos and private isolation that prevail around us. How sad, then, that we have built countless numbers of un-neighborhoods — and continue to do so. We have created luxurious private territory, then totally disregarded the need for cooperative territory.

Often we are shocked by the young who finally withdraw from the air-conditioned, television-saturated environment and perform some hideous antisocial act, completely

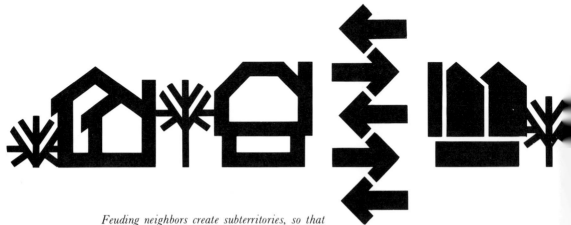

Feuding neighbors create subterritories, so that individuals from houses on one side of the line tend not to socialize with those from houses on the other side of the line.

unconscious of responsibility to anyone other than themselves. We are less often shocked at the great number of marriages dissolved today because this has become a common experience in virtually every family. It is possible that we are not providing an environment that even *allows* for healthy family relationships to exist. Is it possible that the fatherless black families are simply byproducts of a system that has not provided for private territory — thereby negating the need for the male partner in the pair bond?

If we are creating a society that does not provide for, and indeed seems to conflict with, the needs of man, then since man is not easily changed, the society must be restructured for fit; and this restructuring must occur at a comprehensible level. Again, the importance of the neighborhood cannot be overemphasized. It is the building block of planning and the basis for renewal. It provides a real hope for our cities. In the past, when we were part of a hunting group, we chose that group because it improved our chances for survival. This is certainly no less true today. If we are concerned about our children and where we can find the best environment for their security, identity, and stimulation, surely we should choose our neighborhoods with that same cautious concern.

133

The family in the central position physically tends also to be in an equally strong social position.

What should we look for? What comprises a healthy neighborhood environment? What group will my children become part of and grow up with, and with whom will they share the most significant part of their lives? Is there a way of life that can be richer and more rewarding and take them farther? Indeed there is, and it will be found in a small, well-planned, well-organized, inner-directed neighborhood. Let us discuss, then, the critical elements that comprise the healthy neighborhood.

Community Bond

Actually, the pre-World War II neighborhood that most of us over forty can remember just naturally formed a more cohesive community than one would find in suburbia today. The small neighborhood grocery store, drugstore, variety store, and gas station often formed the nucleus of

134

Individuals will make friends with families across a street if the street is narrow and traffic is light.

a community center and were essentially responsible for a more vibrant neighborhood spirit. Healthy neighborhoods, then, are not a wild utopian concept; rather, they have historically on occasion just happened naturally.

We have already discussed the small town and how it naturally developed a pretty good fit. This also occurred in some of our larger cities at the turn of the century, where neighborhoods of first-generation immigrants, with the Catholic parishes as their nuclei, developed into secure and vital communities. Their disadvantage was that they were too narrow, too provincial, and not good long-term solutions. But here we see an excellent application of the formula $A = E + H$. The hazard of the outer community generated the great strength of the homogeneous inner community, and with the parish functioning as the cooperative territory (social space) a very strong community bond was developed.

Individuals will make friends with families to the rear of their home only where foot traffic is generated.

The immigrant settlement is an example of a healthy unplanned community that accidentally has some powerful forces working to its advantage. The kibbutz and commune, on the other hand, are excellent examples of planned efforts to create an ideal community, and certainly a great deal can be learned from them. Again we find that hazard provides the binding force. The kibbutz is faced with the enormous task of transforming a desert into productive agricultural land, certainly a task beyond the capability of any individual. Here we have a dramatic need for community.

The commune is an extremely homogeneous group that has established a definite value system in conflict with the society at large. Conflict is the glue that binds this community together. The kibbutz and the commune are radical examples of communities. Their success or failure again will depend on their "fit"; yet they are creative responses to the profound psychological and biological human need for community.

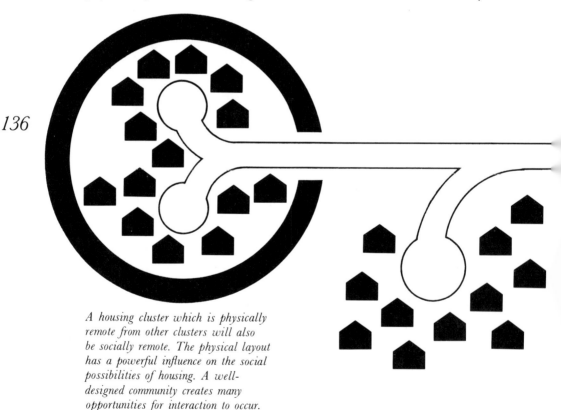

A housing cluster which is physically remote from other clusters will also be socially remote. The physical layout has a powerful influence on the social possibilities of housing. A well-designed community creates many opportunities for interaction to occur.

The church can only serve as a bond when it is geographically part of the neighborhood and when the neighborhood is totally comprised of church members. If this is not the case — and today it almost never is — the church becomes a divisive force which should therefore exist outside the neighborhood...since it cannot promote community bond. This is true of all institutions that limit their membership in any way, for an in-group and out-group situation emerges which is antagonistic to community, and a great deal of energy so desperately needed by the community is diverted into less significant activity. Even the community school that has so much to contribute to the community at large should not exist within the neighborhood because its management is remote, and it draws on an area wider than the neighborhood itself. Institutions with a remote leadership provide programs that often do not solve the unique needs of the neighborhood.

In creating new neighborhoods, or restoring old ones for that matter, what hazard could provide the binding force to create a significant community bond? We need not create an artificial hazard, for our society is awash in subtle and real hazards that unquestionably imperil our survival. Public chaos and private isolation are the hazards. A world of future shock, dominated by unconscious forces creating an anonymous society, often reeling out of control and causing the individual to withdraw inside himself in self-inflicted emotional paralysis, indeed threatens our future survival! This generation has a responsibility to the next generation to create a secure environment in which to grow and learn and mature. Whether we assume this responsibility as a parent or as an adult, we remain duty-bound to respond. Psychologically and biologically we need to be part of a community, and a healthy community must be able to demand a powerful commitment. The community bond, then, must become a kind of secular religion.

Scale

A neighborhood should be physically no larger than thirty minutes' walk from any point to any other point. This is possible in a tract varying from one hundred to three hun-

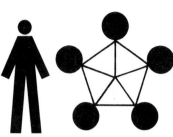

In 1800 neighborhoods were self-sustaining communities; they carried on trade with one another. The individual had power in community affairs.

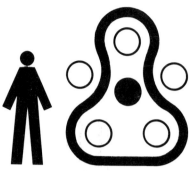

By 1830 one neighborhood began to dominate in each area; it succeeded in annexing the others. The individual had power in the community and limited power in city affairs.

In 1880 the dominant neighborhood had become rich and needed outlying communities to sustain its wealth. The individual had limited power in his community and none in his city.

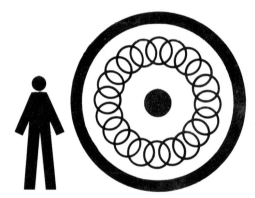

Cities grew, and by 1920 neighborhoods had lost their identity. A power center developed, and the individual was powerless and anonymous in his incomprehensible society.

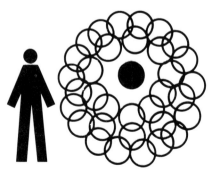

By 1950 the wealth of the city was dwindling; a society without community structure became impossible to govern. The individual fled to the suburbs seeking community but seldom finding it.

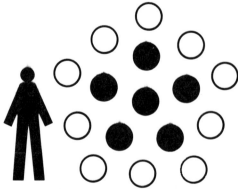

In the 1970s and '80s cities must restructure old neighborhoods and create new ones with identity and political power. The individual must rediscover the significance of community.

dred acres. Its actual physical size is less significant than the number of activities the land can provide. The neighborhood should have a population of two thousand, and should not exceed a population of three thousand. These figures are based upon the capability of neighbors to become acquainted, or at least to know each other by name. Community bond simply cannot exist among neighbors who do not know each other.

Housing should be in smaller clusters of ten to twenty homes, again creating a grouping where people can know each other extremely well, and through this intimate group, individuals can relate to the neighborhood as a whole. This small cluster should have a strong identity, just as the neighborhood itself. Housing clusters should relate well to common areas, and the common open space and pedestrian, nonvehicular areas should be the vital connecting links between the clusters and the community center.

The commons should provide the play and other recreational areas as well as the necessary social spaces. Accessibility of all facilities becomes the main planning objective. A warm, intimate, comprehensible scale is the desired result. The common open space and the community center are where the most significant group action occurs. They should be planned with an active and free atmosphere, allowing for new things to be added in the future.

In the play areas the activities of young and older children should be separated. The moppets should have play areas close to their housing clusters, visually accessible to their parents. This element alone is a more significant luxury than having two baths, if the parents are serious about their offspring. The older children should have play areas that are equally enticing so that they are not tempted to control the younger children's areas. Their requirements, after all, are for larger playing fields, and a great variety of possibilities can occur in scattered locations throughout the open space.

Children should be more vital, intelligent, cooperative human beings for having experienced this kind of environment, and certainly that is the spirit of what we are trying to accomplish in the design. Even if a neighborhood occurs in a densely populated area completely surrounded by miles of development, the open space must project a feeling of openness and not in any way appear restrictive. The open space must reach out for significant vistas and must provide a bicycle and pedestrian pattern of circuitous routes that project a feeling of spaciousness beyond the actual size of the neighborhood.

The community center should be at the crossroads of activity. It should be a natural place for gathering and should provide a great number of excuses for gathering. An athletic club with a swimming pool, a small grocery store along with service shops, a hobby workshop, and a meeting hall are all required to carry on the basic activities of the neighborhood. Higher, denser condominium housing for the older members of the neighborhood as well as the younger married couples who have not yet started their families would be near the community center.

A healthy neighborhood will provide housing for all phases of life, with a wide range of costs. Young people raised in an ideal neighborhood should have the option of settling there with their own spouses rather than having to choose a completely new community. In the same way, the older people should not be forced from the neighborhood because

A young couple buys a home.   *They have children.*

they no longer wish to care for the big houses they needed for raising their families. Security is crucial for both the very young and the very old, and a balanced community should be able to provide it. Also retired people have *time*, and they have a great deal to contribute to community life.

Higher density housing, then, along with the spaces for community services, creates the community center or village square. The parklike atmosphere of the open space and the urban atmosphere of the village square are the two major elements that transform an un-neighborhood with just houses on the street into a self-contained community with tremendous inner strength. The neighborhood is administered and controlled through democratic processes, and the meeting hall is not only symbolic but functional. It also serves as a theater for the local theater groups and as an art gallery. The community must speak out to its people, but its people must be able to speak in return — an aim achieved through local government and local cultural expression.

The hobby workshop provides a great variety of woodworking equipment as well as kilns for pottery, and so forth. Local crafts are extremely significant to a vital community. The astronomical observatory, though controlled by adults, is used extensively by the young. The cooperative

The children move out.

Neighborhoods in which housing is all of a single type are self-destructive. As residents age, the neighborhood ages. There is no infusion of new ideas, no recycling of homes, no chance for positive transition.

nursery, where education is stressed, is a necessary element in the community. The athletic club with exercise rooms, saunas, and a swimming pool for year-round swimming not only provides for exercise but also offers meeting rooms and other social spaces that duplicate many functions of a country club. Everyone in the neighborhood is a member; and again, the stress is on family participation.

No organization that is exclusive in any way can exist in the community. No organization strongly controlled from the outside can exist in the community. In-group/out-group formation can destroy the community bond. For this reason, city or county government should not have control of the open space. There is always the temptation to

A healthy neighborhood provides housing for all phases of life over a wide range of economic capabilities. There is provision for renewal through interchange among young and old, struggling and affluent.

delegate the costly task of maintenance to the city parks department, but this is a mistake. The neighborhood home-owners' association must not only maintain its own open space but also must own and maintain all of its own community facilities, including the athletic club, service shops, grocery store, and so forth. The money earned through rental on these spaces will not only provide operating capital but will create a substantial fortune in real estate through the years. It is hard to imagine a neighborhood deteriorating that yearly grows more wealthy.

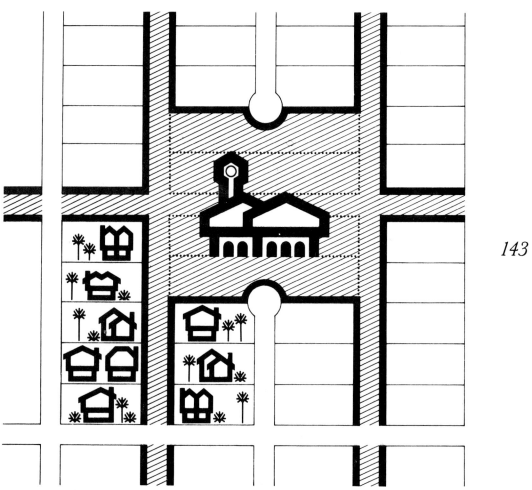

Un-neighborhoods can be remade into neighborhoods by buying four corner houses, creating two cul-de-sac streets, building a community center, and establishing some open space throughout.

Conclusions

In the last hundred years, the parent-child relationship has changed considerably, and not always for good reasons. Children in an agrarian society are naturally an asset because they can help with the enormous amount of work required on a farm. As America has become more urbanized, the child has slowly become not an asset but a luxury.

What has really happened is that we are postponing the time when our children become "assets." We need the young to properly take over the duties we are now performing, and the future quality of life is being determined now by the way we train our young. Parents are just as important today as they have ever been. Earlier we discussed the need for the pair bond to provide the young with security and education while they mature, and this remains the parents' task.

As parents we have been deluded. We have entrusted our security to a police force that is remote and usually inadequately staffed, and in the same way we tend to abdicate our own responsibilities in turning over the education of our children to an institution that again is both remote and usually understaffed. This is not to demean public police protection or public education. On the contrary, both are extremely important, yet they are limited in what they can do. We might go so far as to say that they were never meant to do many of the things we expect from them and falsely believe we are receiving. A well-organized neighborhood can attack the dope problem, for instance, with considerably more muscle than can individual parents. A neighborhood night watch can successfully reduce the threat of burglary. Parents, dissatisfied with the education system, can organize supplementary enrichment courses along with special tutoring services for students with problems. The neighborhood dedicated to providing a good educational environment is certainly going to provide the basis for an incredibly high aspirational level among its children.

We can afford more, we can accomplish more cooperatively than we can individually. Perhaps the rebellion of the young is a recognition by them that they have not received the crucial elements of security and education that only

A nuclear family is a group composed of parents and children inhabiting a private territory.

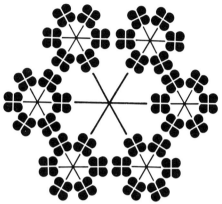

A micro neighborhood is a group of ten to twelve families who share a defined area and some limited amenities.

A neighborhood is a group of ten to twelve micro neighborhoods (100 to 125 families) who share a defined area and a variety of open spaces.

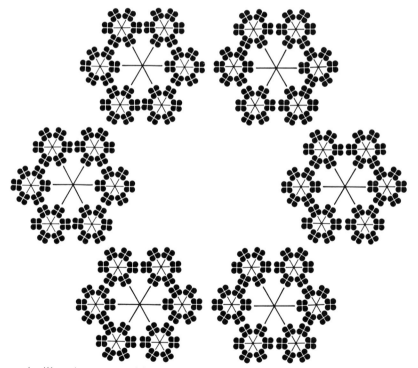

145

A village is a group of five to six neighborhoods which function as a small town. It is governed by a village council and administered by a professional administrator.

parents in a healthy community can give. Perhaps they had parents who really tried; but it is likely that the community from which they were also to receive significant elements of security and education was such a nonentity that the parents alone were not equal to the task. The point is that parents must be dedicated to their task of raising their young, and they must diligently seek out a community in which they will have as many advantages as possible in fulfilling their responsibilities. No outside adult can ever have the influence on a child that his parents have, and no school can ever have the impact on a child's life that his community has.

In the United States we delude ourselves that the parent can solve the significant problems alone; but in some of the more radical social experiments, such as the kibbutz and commune, the community rather than the parent is entrusted with the children. Both solutions are only half right and therefore all wrong. The combination of good parents and a healthy community is the only answer. It is the answer that

A micro neighborhood results when eight to twelve families give their defined territory a name, a focus (such as a playground), a function, rules, and leadership.

is deeply imprinted in our genetic inheritance. It is the way of the past. It will remain the way of the future for a long time.

Many sociologists will argue that homogeneity — that is, common characteristics, interests, and background of inhabitants — is critical for healthy neighborhoods. They further argue that racial, economic, and religious differences can inhibit the formations of the healthy community. This is of course a very real problem and one that must be resolved. If the members of the community are all committed to providing a healthy environment for their children, then this higher goal would create a new homogeneity which will override the racial, economic, and religious differences. After all, a community successfully producing healthy, vital, intelligent, cooperative young adults will be able to endure, regardless of its petty tensions and divisions. Indeed a healthy, integrated neighborhood makes an enormous contribution to the quality of life of the entire nation, but more importantly, it is a fulfillment of the nation's highest purpose.

147

A true neighborhood results when 80 to 100 families give their defined territory a name, a focus (such as a clubhouse), a function, rules, and leadership.

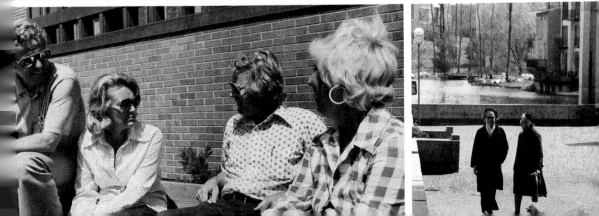

CHAPTER NINE

Educational vs Pseudo Communities

"The thing that's obsolete is the way we've been managing the nuclear family. Two people can't be all things to each other all the time. Our only hope for successful marriages is to build stronger collectivities — groups of people living together with many mutual interests. We need cohesive, three generational communities where each person, though married, can draw on a great number of people around him, rather than just on his spouse. This will filter down to the kids as well."

Margaret Mead

A valid community provides a secure place to educate the young for survival in a world with many hazards. This is true of animal communities, primitive human communities that have lasted for millennia, and agricultural communities that are so much a part of our recent past, and remains true of communities formed in a modern technological society.

The problem remains the same: The demands of basic human needs must ultimately be satisfied to achieve a successful societal structure. If this doesn't occur, we have a breakdown of society. Even animal communities must solve the needs of the members of the group in unique and special ways. We often think that basic social patterns of a species would remain universally the same. This is not so, for as environmental conditions change, a chimpanzee community will respond directly to the immediate environmental requirements and will vary considerably in terms of territorial requirements, pair bond, and basic community structure. An arid desert and a dense jungle are environments with totally different food-gathering requirements. This of course has immense impact on the basic social fabric which must respond to the functional requirements of the specific environment. The end result, however, is always the same. Successful animal communities must provide the young with security as well as teach them the harsh realities and danger of the natural environment in order to ensure the survival of the species.

In the same way primitive societies of man vary considerably. Sometimes the family takes precedence over the community. Sometimes the community at large overshadows the family structure. This is often common in hunting and fishing communities when the livelihood depends upon the intense cooperation of a large group. But again the binding force of these great varieties of community structures remains the same — that of providing a secure condition and then providing the education for survival. Only then does the community fulfill its vital function, only then does it become the immensely significant survival mechanism it was meant to be.

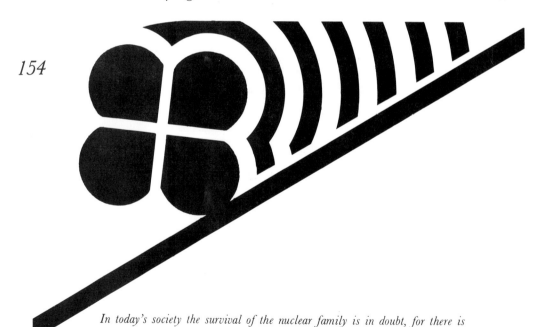

In today's society the survival of the nuclear family is in doubt, for there is no community structure to support it and no societal foundations to encourage the neighborhood community.

Because of *Homo sapiens'* upright position, with the resulting pelvic structure which limits the size of offspring on the one hand and with the very large brain that we require on the other hand, our young are born with a mere 25 percent of their brain capacity. This is extremely low compared to the chimpanzee who is born with 65 percent capacity. The problem is further compounded by the slow maturation of the human compared to the chimp who goes through puberty at eight or nine years of age, as opposed to twelve to fourteen years for humans. The human baby is helpless at birth and remains in need of considerable care for a long time. The pair bond is of course the resulting survival mechanism, and the community is essential to the proper functioning of the pair bond and the education of children.

One is overwhelmed by the awesome reality that just possibly contemporary man has lost his direction and seems no longer concerned with providing a secure environment for the young, nor is he willing to put forth the effort to educate the young, but rather is hell bent on securing only the desires and needs of the adult community who are able

A nuclear family, dedicated to the security and education of the young, needs the support of an identifiable community, plus the foundation of a society dedicated to the community ideal.

to pay for it. Think of the number of golf courses built as opposed to the number of playgrounds. Think of the money spent on new automobiles and air conditioning as opposed to money spent for the institutional education of the young.

As the adult generation stands in astonishment and even shock at a youth culture committed to a life-style void of the value system of its parents, it is possible the extremely busy parents did not have the time or inclination to transmit this value system. In the absense of any direction, the young often indulge in excessive experiments in pair, group, and community relationships. College professors are beginning to recommend unfounded alternate solutions to the nuclear family, disregarding an amazing record of failure of these experiments. On the other hand, many adults are retreating into sterile, all-adult, pseudo communities that can never become real communities because they are without purpose. Again the main function of a community is to provide a secure place to educate the young for survival and a great deal more.

The nuclear family is not an outmoded social structure; rather, an anonymous hostile environment has greatly inhibited its proper functioning. We cannot and perhaps never will be able to mold human needs to the demands of an arbitrary social structure. We must accept basic human needs and their satisfaction as functional requirements of society and therefore must create a social fabric that is not antagonistic to the nuclear family.

Societal pressures and current community structures steer adults and children in the opposite directions.

It is useless to glorify the primitive society beyond the bounds of reality as the ultimate ideal. Certainly as these societies evolved, they often achieved an incredible degree of "fit" that has lasted for centuries, and are therefore worthy of study.

The Wogeo Culture of New Guinea

Ian Hogbin, an eminent anthropologist, has given us an excellent description of the Wogeo culture of New Guinea. He portrays the everyday life of a small, technically simple, preliterate society and gives some specific examples of how the child relates to the family and community at large. As soon as the child is weaned, it takes a significant place in the community and is treated and accepted as an adult at this

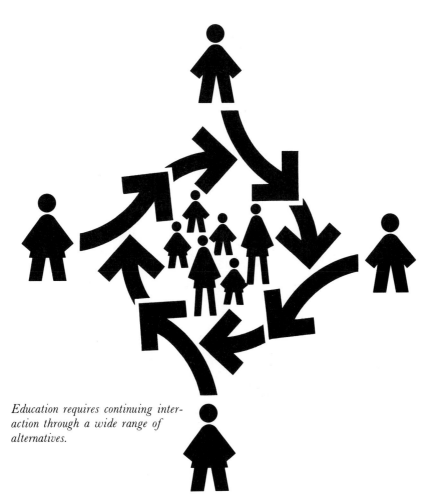

157

Education requires continuing inter-action through a wide range of alternatives.

early age. Often the child is assigned an area in the family garden and becomes responsible for a simple crop. The child may also be given the added responsibility of feeding and caring for several young pigs. When one of the pigs must be killed for food, the father first consults the child. Adults do not separate the children from essentially adult activities; rather, the children's presence is both accepted and expected.

Even during ritual dances, the children perform alongside the principals, doing their best to mimic. Seldom is force used in disciplining the children; rather, lengthy explanations serve as a method of teaching a moral value — and also as a mental punishment. (Young, active children would rather do anything in preference to hearing a long dissertation on their mistakes.) The children have the rare luxury of being needed, and at an early age are relied upon to help complete the daily chores. Even the very young play at work, for that is how they learn. When children are asked why they do things certain ways, the answer almost universally is: "That is our custom," which is a result of a natural way of transferring the value system while parents and children are working side by side.

Education is not haphazard; it is a serious effort of the parents, relatives, and the community at large. For example, the boys in the village build model boats and float them on a pond. The village elders then gather and criticize the boys' efforts in terms of the length of outriggers, the shape of the hull, and so forth. This direct kind of education has positive results, for the boys learn from the critique of their elders the things they have done right and those they have done wrong.

Relatives play an important part in the education of the young, but the father is primarily responsible for the sons and the mother for the daughters. A poignant story is told of a young boy collecting nuts high in a tree. He somehow got out on a limb and, feeling incapable of getting back to the trunk, cried out to his father who was collecting nuts in an adjoining tree. The father immediately called instructions of how to get back to the trunk, and then a rescue was completed with the aid of a rope. On the ground the father comforted the boy for about an hour and then suggested they climb together. "If you don't climb the tree now, you may be frightened tomorrow. I'll follow close behind and tell you where to put your feet."

The parents have a strong conviction that their way of life is a kind of "sacred" truth, and their children must be taught to be committed participants in this way of life. Here we see a community undoubtedly centuries old that in a very simple, direct way solves basic human needs with enormous success. Here we find committed parents living in a vital community totally dedicated to the education of their young — not for survival alone, but also to perpetuate their "sacred" way of life. Modern man has much to learn from these primitive people.

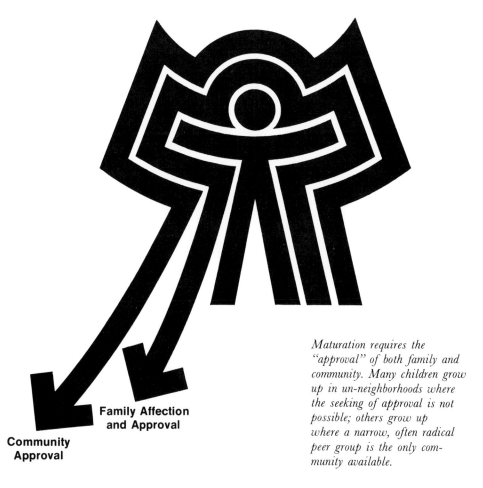

Family Affection and Approval

Community Approval

Maturation requires the "approval" of both family and community. Many children grow up in un-neighborhoods where the seeking of approval is not possible; others grow up where a narrow, often radical peer group is the only community available.

The Iks of Uganda

Now let us consider another primitive society, the Iks of Uganda, brilliantly described by Colin Turnbull in his book *The Mountain People*. Here we find a people who once were a vital hunting community, where the pair bond was subordinate to the larger community structure, yet a very successful community existed. The requirement of intense cooperation for hunting in an extremely dry, barren environment established the societal form. Recently the government established their hunting area as a game preserve, which totally destroyed their food-gathering pattern and ultimately the entire societal structure. No longer was there need for intense cooperation; the hazard required for a strong community was gone. All the outward manifestations of a behavioral sink began to emerge. The maternal instinct deteriorated — children were put out at the age of two or three to forage for food on their own or in hunting bands of young children. The father no longer brought food home to his family but simply took care of his own needs. Adults allowed their aged parents to starve even when they had more than enough food personally. An old blind woman falling down a hill and a small baby crawling into a fire would illicit laughter but never help from adults.

Turnbull points out that for the time being, the Iks are surviving as individuals but at an incredible cost. He says:

> They have made of a world that was alive a world that is dead, a cold dispassionate world that is without ugliness because it is without beauty, without hate because it is without love and is without any realization of truth even, because it simply is.

The chilling part of the book is not his description of the Iks, but rather his concern of modern man:

> ...and the symptoms of change in our own society indicate we are going in the same direction...family, economy, government, and religion, the basic categories of social activity and behavior despite our thinking are no longer structural in a way that makes them compatible with each other or with us, for they are no longer structural in such a way as to create a sense

of social unity involving a shared and mutual responsibility between all members of our society.

We have for far too long placed our trust in antiquated institutions that have not solved society's problems.

The Confusion of Today

We have discussed the merits of a healthy primitive society and we have also seen how it is not immune to collapse. Any society, if it is not able to respond to new environmental demands, ultimately disintegrates.

We have much in common with primitive man. We have limitations. We possibly can survive as individuals (for a limited time) but only in a hollow, meaningless way. Communities dedicated to providing security and education for the young are the only answer. This is the limitation that nature has placed upon us, and it must be accepted and respected.

Communities in the past evolved slowly without too many external influences, and often they achieved a high level of fit. It is only recently, when we have attempted to create instant communities, that we have developed some monumental problems. Men who were given far more power than they deserved disregarded or were ignorant of the broadly based needs of a community, and human storage units clustered in un-neighborhoods are the tragic result.

The fact that this was wrong from the beginning was evident, but this did not deter us from repeating the same mistakes again and again for the last 35 years. We now have men touting the concept of community, but what kind of community are they providing? In far too many cases they are only perpetrating another fraud. Recreation is the answer for some — a quiet, private (incredibly dead) world away from children is another. But neither of these elements can support a real community.

Recreational or Educational Communities?

Is it crucial that we develop a generation of golfers, swimmers, tennis players, or skiers? Is total recreation an urgent element for the young? What kind of adult will result from this kind of pseudo community?

The main reason for criticizing the recreational community is that it is being promoted as a kind of ultimate answer. The great majority of new community planning has been based on this format. It is not that recreation is bad per se, it is simply that it is a shallow solution. Worse yet, some of the problems emerging from recreational communities are frightening. For example, the wife plays golf with a women's golf group, the father golfs with his group, and the kids take swimming lessons with others of their own age. Instead of the community amenities bringing the family together, they are literally tearing it apart, as witnessed by a high divorce rate in some new communities. Many times athletics become excessively competitive and often dishearten the less gifted athlete, leaving him feeling insecure and left out. Competition is the antithesis of community. Cooperation is the basis of

162

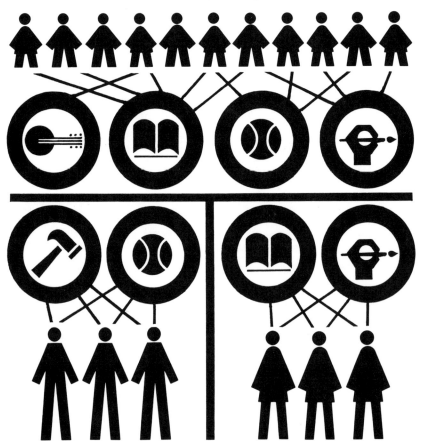

An un-neighborhood separates men, women, and children with its amenities, its services, and its institutions.

community. Certainly there are some survival lessons to be gained from athletics, but they are highly overrated. Educating the young to live and participate in a healthy society is the main function for community. Recreation alone is too limited a base.

Why not create a generation of workers, builders, creators? Again the strong emphasis on recreation is not primarily aimed at solving the needs of the young, but rather those of the adults. Adults need a break from their daily work, but children need a break from continuous play, and also need to learn how to work and become skilled in basic survival endeavors — such as growing crops and caring for livestock, or building simple structures, or taking care of simple house and automobile repairs. But beyond this, a child who shows a particular talent in art, music, writing, and the like ought to receive encouragement and help, not only from his parents but from other adult members of the community who also have the talent. A one-to-one teaching relationship can never be surpassed.

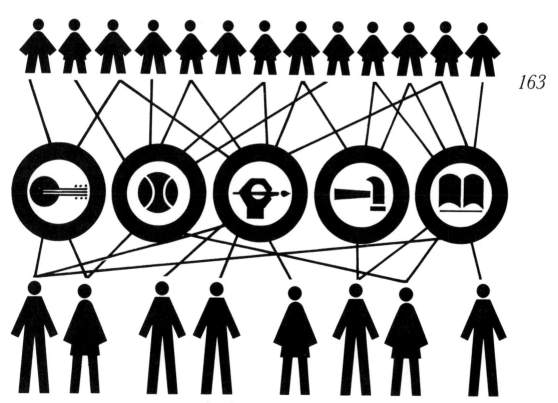

An educational community provides for the interaction of adults and children through a wide range of educational amenities and services.

Our educational institutions attempt to provide a broadly based education, but their facilities and staff are generally woefully inadequate. More importantly, they remain a remote, impersonal institution, operating at a scale that would almost always limit their real effectiveness.

Professional educators today are finding that the Soviet Union and the Republic of China are doing a substantially effective job of training their young to live in a highly structured socialist society. A national policy of "behavioral engineering," where individual effort is totally subordinate to group achievement, is indeed repugnant to Western values; yet we cannot deny its effectiveness in outwardly achieving "well-adjusted," good citizens.

The same educators are finding that the child growing up in the United States is not receiving adequate training for the society he will ultimately inherit and contribute to.

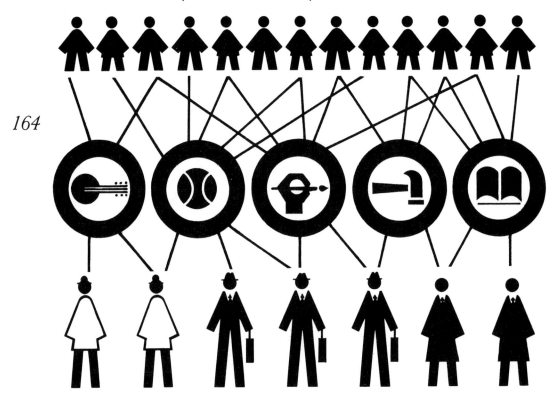

A pseudo community limits the types of interaction between children and adults by restricting the kinds of professionals, craftsmen, and viewpoints available.

In Urie Bronfenbrenner's *The Two Worlds of Childhood*, he describes our ineffective, fragmented process of education:

> The primary function of schools became the teaching of subject matter. Character education or what the Russians call "vospitanie" was left to the family and the church. The role of the church in moral education has withered to a pallid weekly session at Sunday School.

He further maintains that the parents have also become ineffective in an age-segregated social order, void of community, and then concludes:

> The vacuum, moral and emotional, created by this state of affairs is then filled by default — on the one hand by the television screen with its daily message of commercialism and violence, and on the other hand by the socially isolated, age-graded peer group, with its impulsive search for thrills and its limited capacities as a humanizing agent.

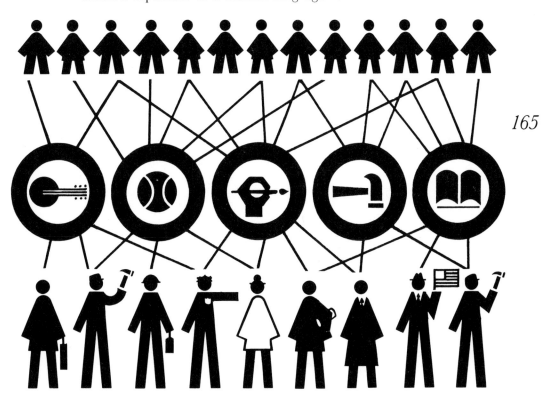

An educational community provides a wide range of contacts between children and adults through amenities and services designed to deal with the interests and needs of both.

Television, often referred to as "the third parent," and the peer group then become the significant elements in the child's life...a grim similarity to the Iks, who turn their children out at a very early age to fend for themselves.

In William Golding's *Lord of the Flies*, a pre-adolescent group of boys marooned on an island epitomized the ultimate in peer power. "Piggy," who somehow expresses the best in civilized human relationships, is finally killed by the uncontrolled sadism of the other boys. When rescuers finally arrive too late to stop the brutality, their first question is: "Are there any adults — any grownups with you?" Children do not naturally find their way in the world; rather, they are in dire need of adult help and direction.

The Soviet solution certainly is not satisfactory, but it is vastly superior to the totally nonstructured environment that our children are thrust into. Bronfenbrenner says, "If adults do not once again become involved in the lives of children, there is trouble ahead for American society," and concludes:

> We are therefore faced with the necessity of developing a new style of socialization, one that will correct the inadequacies of our contemporary pattern of living as it is affecting our children and provide them with the opportunities for the humanizing experience of which they are now bereft.

Parents must be committed to their children, and they must further search out and become a part of a community that is also dedicated to the security and education of the

166

When education is a horizontal function, the influences on a child consist only of parents, peer group, and perhaps the limited viewpoint of television.

young. Thus the function of the small community takes on inestimable significance. It is the critical survival mechanism through which the value system of society is perpetuated. School and church must reinforce this process, but we must recognize the limitations of these remote institutions as contributions only, and not the community process itself.

All Adult Communities

Can the older generation achieve the respect necessary for a satisfying old age if it is segregated and its members become nonparticipants in this most significant process of educating the young? On the other hand, who has more time,

When education is an ambilateral function, the influences
on a child include parents, peer group, other adults,
and the reality of "community."

more wisdom and common sense to share? The closer the generations are, the more competition there is between them. The older generation then inherits that rare stance of being above the battle. This strategic position provides not only a great opportunity to referee the foray, it is a necessary function of a well-structured social fabric.

The need for the older generation has not diminished; rather, our compact houses and our un-communities have often simply squeezed these people out. Clever developers have seen the problem as has everyone else, but they

168

The aged, relieved of the responsibilities for running society, have a critical responsibility to teach and influence the young so they can move into the central position in adult society.

have rationalized it into a great business opportunity. So-called "empty nester" senior citizen communities have emerged which have no right to the name "community," but nevertheless are as lucrative as they are banal.

This unnatural pseudo community of humans tries to establish a community bond among people who don't really need each other. They need someone their own age to talk to and share their problems with, but too often they can't really assist one another. The young, on the other hand, need the old — to read to them, to watch over them and take them on excursions when their mothers are too busy. The old often need help in maintaining their home as well as direct physical help that their children, who are now parents, don't have time to perform. Again we see a powerful excuse for this alliance of alternate generations, but the alliance should function as an integral part of the community bond that is the binding force in a natural multigenerational, authentic community.

Judging from recent findings of geriatic studies on aging, we see that perhaps the simple element of being a significant and respected participant in society does more to prolong life than any other element, and there is a huge, confused, and often lonely younger generation who desperately needs their help.

The Educational Community

Modern man is developing an illusion that since machines are going to do the work anyway, he will have to find other ways to spend his time. A machine cannot paint a painting, build a stone wall, or invent other machines. In fact, far too often our machine-made environment is not esthetically pleasing, and therefore not emotionally satisfying. We are still blessed with need for many hand-made objects, and if we are to create an ideal environment, nothing is more emotionally satisfying than to be personally involved in the creative process. Consuming can never give the degree of pleasure that creating can. A community whose amenities are based primarily on consuming is a shallow excuse for a community. An individual who is primarily a consumer is a poor excuse for a human being.

As we discussed earlier with respect to chimpanzee communities, the environment required certain food-gathering techniques which proved to be the catalyst for the social fabric that finally emerged. In that same sense the kinds of amenities provided by the community will be the catalyst for the kind of social fabric that eventually emerges. The amenities will also be the "behavioral engineering" structure through which the adults will educate and influence the young. With this in mind, community amenities take on an entirely new significance.

An educational community would require essentially the same amount of space as a recreational community, but instead of dedicating a hundred acres to a golf course, the land would be used for a co-op farm where families, particularly the children, would receive a gardening assignment. Some might even have a cow to milk, a pig to feed, or chicken to care for. Where space is restricted, fruit trees could be planted in the parking strips and recreational space to provide the community with a crop. The educational benefit of being personally involved with the food-gathering process should be apparent. In our highly interdependent society, a huge segment of our population grows up with but a fuzzy concept of the food-gathering process. To feel the satisfaction gained

170

Institutional education is characterized by the need for one adult to relate to large numbers of children. While important to society, it is highly structured and transitional in nature.

from a successful crop or the anguish from a crop failure is an educational imperative. To be able to see first-hand the simple ecocycle of grazing livestock, planted fields, and crop rotation should be part of the educational heritage of our children, if they are to solve the ecological problems of the 21st century.

The open spaces for playing fields, the tot playgrounds, the creative play areas — all should be lush educational environments. The village square should provide the community focus for a vital new life-style. Community government, carried on in the village cluster, should have many participants, which would include all of the young. Here they would learn directly how a community functions and how each individual must be part of and contribute to a society if that social order is to remain healthy. Again, here is a crucial survival mechanism — far too many have grown up without a direct personal involvement. School government makes a half-hearted attempt, but school government is not real government. How can we continue to be a democracy if the young are seldom or never personally involved in democratic processes?

The craft shops with weaving, pottery, and woodwork, the village theater group, the village rock band and string quartet are all activities that can raise the cultural

Community education is more likely to exhibit a one-to-one ratio between adult and child. It continues throughout a lifetime, and is possibly more significant than institutional education.

life of a community to a superior level. When we create something beautiful, it is immensely satisfying personally, but we also share it with others. Finally we have created the optimum condition of providing a place for the creative person to give and a community to respond. In modern society, this has been a luxury that far too few creative people and far too limited a community have been able to enjoy.

A local industry which manufactures products that are sold on the market would be another significant element. Ideally, people should not have to commute a great distance to work; but more importantly, children should have an opportunity to be closely associated with and finally to actively participate in the process of production. It is an exciting, invigorating human drama that should not be isolated from daily living. It would of course have to be a clean industry, but there are many possibilities.

The child growing up in a community where he is an active participant in local government, where the arts and crafts are an integral part of his everyday experience, where he contributes personally to the food-gathering process, and where a local industry gives his community an identity (because of its quality products) in the nation would have little excuse for not succeeding personally. Given the critical advantage of having concerned parents, the child has had the added advantage of being intensely involved with a great number and variety of other adults who are totally committed to making a simple, comprehensible society function properly. The educational possibilities of this type of community are overwhelming.

Specialists in child development tell parents what they are doing wrong and that the child needs a parent who is also a psychologist. This argument infers that parents are either void of natural maternal and paternal instincts and intuition, or that the instinct and intuition are wrong. What really is wrong is that we are creating a life-style that inhibits the natural functioning of the nuclear family. Instead of becoming a father-confessor-psychologist for your child, which seems to be rather forced, it would seem that an alternate, more natural solution must be found; and one answer is living in a situation where adults other than the parents can help share the burden. It is possible that the significant determinant of a child's education is not the quality of institutional education he was exposed to, but the quality of adults surrounding him as he was growing up.

Ultimately we must ask ourselves: Is our life-style really worth perpetuating? Are remote and impersonal institutions directing a life-style from such a great distance that we are becoming lukewarm participants in a poorly functioning process? The way we are going...is it a vital survival mechanism? If it is not, then we must personally seek out or help create a life-style for which we have conviction, and with this conviction will come the natural determination to share it with the young. Only a "sacred" way of life is perpetuated generation after generation, and the small educational community is the ultimate vehicle.

CHAPTER TEN

A Day in the Village

W e have discussed the theory of the ideal neighborhood, but now let us get down to specifics. Let us examine a family and its relationship to its neighborhood (of course we are assuming an ideal model in order to more graphically portray the new and vital life-style).

Our typical village — typical in the sense that its concepts are being used in one form or another even now — was begun about four years ago and is an actively functioning neighborhood. Completed just last year, it has a population of three thousand people. The developer responsible for the project recently turned over the last remaining properties and legal rights to the homeowners' association, which is controlled by a board of directors and administered by a professional manager. Now the village with all its amenities is completely owned and controlled by its inhabitants.

Don, an advertising executive, is one of the duly elected directors, and he and his entire family are deeply immersed in the activities of the village. An average day's activity should give considerable insight into the quality of community life that they experience. Don and his wife Jane have three children: Mark (fifteen), Jill (twelve), and Steven (three). Jane's widowed mother Mary also lives in the village in one of the condominium units at the village square.

Don has found the change of pace in the community a relaxing diversion from the hectic nature of his work. Jane has found self-realization in the activities of the community. The children have possibly benefited the most because they have plenty of time for the great variety of activities available.

We find them on a Saturday morning, preparing for a reasonably active day. Jane has just ripped through the house, getting things in order, which is not an overwhelming task since having most of the furniture built in reduces housework immensely. Mark and Jill come sliding down the fireman's pole from the loft (where they often camp out on weekends) and proceed to pitch in and help. They are personally responsible for their own rooms, which sometimes develops into a fair-sized task. Their rooms are divided into two areas: beds, desks, bookcases; hobby display areas, and the hobby closets. Each one's hobby closet is in constant use, and often a particularly engrossing project somehow spills out into the room. Many of the houses in the village have such closets... the idea being to provide an area to work on projects that would not have to be cleaned up each time. The idea has possibly worked too well, for often the hobby closet gets so much attention that the rest of the room suffers. But then, what better use could the room have?

Steve, the three-year-old, is grumping around because the co-op nursery is closed on Saturday and he will not see all of his friends. On this particular Saturday his grandmother (Mary) is going to take care of him, and even though he likes to complain a little, he thoroughly enjoys a day with her.

Don has already finished the little bit of yard work required and is just eating his breakfast. His plan for the morning is to go down to the hobby workshop and work on the dining room set he is building. He has already completed some built-in furniture for the playrooms and Steven's room, as well as a couch and coffee table for their living room. (The village designers also provided a great variety of extra furniture plans so that people could build many pieces for themselves, and Don was one of the first to take advantage of this.) He experienced deep satisfaction in being able to provide in a personal way for some of his family's needs. He also enjoys creating something himself. In fact this time, although there are some excellent designs for dining room tables and chairs, he has designed his own, and regardless of what anyone says, he is sure that at least *he* is going to like them.

Jane is an active member of the craft guild and has been most involved in weaving. Today at four o'clock a guild show is opening at the athletic club. Normally it would have been held in the community center, but there is also a play tonight so the locale had to be changed. She is going to the craft center with Don to finish off a few things and help others get their items ready. She plans to enter three pillows, a shawl, and her real pride, a large tapestry she worked on for an entire year.

Teen-aged Mark has to attend his final play practice before the actual performance this evening. The play is a musical, and although he does not have a substantial part, the whole family has been forced to become familiar with all the songs. After rehearsal he will check the playground area that he and his group maintain. He is in charge this month, and although he cannot be there to help with the work, he will still hold the final responsibility.

All the grounds and common spaces are maintained by the youth organization, which is structured like a miniature government. They receive final instructions from the professional manager, and the actual work is administered and carried out by the young people themselves, organized in teams of essentially the same age group. Here in the neighborhood the children are not only well cared for, they also fill a crucial need, and it is amazing what a monumental difference this has made! There are few community discipline problems, and those few are handled by the youth court, whose decisions are not renowned for their leniency.

After checking the playground, Mark hopes to have some time to spend with Don at the craft center, for he too has become interested in woodwork and has completed some fairly complex projects.

Jill, who is a good swimmer, will attend her swimming class first. Afterward, she and the girls' group to which she belongs will carry out their assignment to help an elderly woman who lives in the condominium. Their specific job today is to wash the kitchen walls and windows. Jill and her friends did not complain about this assignment because they have helped this woman before. Although the elderly woman has been troubled with arthritis and cannot do heavy work, she is an excellent cook, and there is always an abundance of cookies, cake, and candy.

Finally Mary arrives and enjoys coffee with her daughter before the grandmother and Steve depart for their day's activities. Steve knows what to expect, and he is getting his tricycle ready for a long, long ride. He likes the play area directly behind his house, but it is fun to visit the five or six other "tot lots" they will pass on their morning walk.

The village has a series of narrow, landscaped parks that provide pedestrian walks and bicycle paths throughout the project. Although the village comprises a mere one hundred acres, the circuitous route through the narrow park conveys a real sense of spaciousness. If Steve and Mary move reasonably fast, it will take virtually the entire morning to walk through the network, with a few minutes here and there to enjoy the small playgrounds.

There are larger open areas along the stream where the boys play sandlot football and the girls play tennis. The youngsters plan to have a trampoline soon. There are no organized teams at the village because Little League teams are available outside the neighborhood; the emphasis here is on individual achievement. It was decided too that the casual games kids play for sheer enjoyment are a much more vital learning experience than highly organized athletics with adult referees making all the decisions. There is in such thinking a real effort not to allow in-group and out-group situations to develop.

Steve and Mary will also pass the creative playground, a fenced-off area full of brick, scrap lumber, and a variety of other building materials the kids use for creating their own environment. It is constantly being worked on, with some amazing things being built. A real shantytown has emerged that has achieved a kind of ramshackle magnificence.

Today is a refreshing break for Mary. During the week she is busy at the village gift shop in the daytime, and in the evenings she teaches pottery at the craft center. Since she is an exceedingly good potter, her only regret is that she finds so little time to do work of her own. On the other hand, along with many others her age, she finds life in the village most satisfying. The retired people often find that in a real sense, they are not retired because there are so many opportunities not just to serve, but places where they are really needed.

At noon all the family are back at the house for lunch except for Steve and Mary. Mary is meeting some friends for lunch at the delicatessen at the square and, like all grandmothers, she wants to show off little Steve. Lunch, as usual in the summertime, is out on the patio. The house has a very small outside space, but it is beautifully landscaped and serves as an excellent summer living space. There in the shade of a large maple tree, with a magnificent view of the green park and the mountains beyond, the family enjoys idle chatter and a delicious lunch together.

Don and Mark have a Saturday afternoon tennis-playing ritual. After tennis they try to find time for a bicycle ride. Cycling is always popular in the village, and the cycling paths are already beginning to need repair. Every Fourth of July a six-mile bicycle race is held as part of the big celebration. Mark took first place two years ago, and has tried hard to keep in shape for the forthcoming race.

If they move right along they will have time for a swim and a sauna at the athletic club. Here they have a beautiful indoor pool used year-round, along with exercise rooms, a ping-pong and pool room, meeting rooms, and so forth. The club has been so successful that an addition to the clubhouse is being planned, and there is a new outdoor

Olympic-size pool that was just completed this spring. The older kids have full responsibility for maintaining the place, so the younger ones do not dare mess it up. Three years ago a couple of young creative artists were discovered engraving graffiti on the bathroom walls. They were summarily tried by the youth court and sentenced to one year of cleaning toilets. Word got out, and since then the walls have been virtually graffiti-free.

Jill is meeting a friend at the craft center where she plans to finish a basket she is weaving. The best will be in today's craft show. Jill and her friends have waited until the last minute of course, and the pressure is on. If they finish in time, they plan to spend the rest of the afternoon tubing on the canal. The canal runs the entire length of the neighborhood, and even though the water is quite muddy and undoubtedly somewhat polluted, it is a great place for tubing. Huge water fights often develop, and occasionally when things really get out of hand, adults who are particularly known to the kids have been thrown in the canal, good clothes and all!

Jane is at her craft guild meeting taking care of the last details before the show opens. She has a real fondness for the other people in the guild. They are all willing to share new methods and techniques, and they have gained real pride not only in work of their own but in the work of the entire guild.

Mary and Steve will visit the afternoon puppet shows that are given in the commons by some of Jill's friends. The kids have to sign up for the stage in advance, so they try to put on a fairly good show. The show's quality covers a wide range from exceptionally good to extremely bad, but then what can you expect from experimental theater (not to mention that the kids Steve's age enjoy them all equally).

It is four o'clock and people are beginning to arrive for the craft show. Don greets Jane and they walk through the building together. Don comments that the weaving is extremely good, much better than it was last year, and the pottery is good, but the basket weaving is not up to last year's standard. The woman who had taught basket weaving moved away last year. The kids were still inspired by what she had taught, but they really needed direction.

When they come to the woodworking display, Don is overwhelmed. Here is an oak dining room set that is magnificent. He studies it for a long time. After looking at the other chairs, coffee tables, and couches, he confides to Jane that he would not dare show his dining room set in this show next year. He is going to have to restudy his entire design, and this show has taught him many things.

At the punch bowl Don greets another member of the governing board, and they chat a minute about a local problem. A young girl from one of the less affluent families has won a partial scholarship to a prestigious university, but her parents still cannot afford to send her, so their neighborhood is working out a way to secure and back a loan for the young student.

A fairly large number of the neighborhood's young people go away to school, and often to good but expensive schools. Others are interested in learning a trade, and the best trade schools are sought out. The village feels a neighborhood obligation to somehow help to provide an education for every young person who sincerely wants one. After all, the entire purpose of the community is based upon the idea of a vital place to educate the young in a secure environment — and the young are the end product of the process. The parents can often afford to help their children through school, but the neighborhood is always there to back them up.

185

Mark walks through the show with several friends, and all are impressed by the work of their contemporaries. Jill is just a little disappointed in her work and almost wishes she had not entered. Don assures her that her basket is certainly worthy of showing.

Many of the good items have already sold, and as in any display of art, some of the best things are not for sale. Jane would not think of parting with her tapestry. The living room already looks naked since that tapestry was removed for the show, and it had been hanging there for only a week.

Art, whether it is painting, sculpture, crafts, music, dance, or theater, is given strong emphasis here in the village. There are at least four painting and sculpture shows and two craft shows each year, and at least one play is always in production — not to mention an almost unbearable number of musical recitals and several ballet productions. The young people are infatuated with folk dancing, and although they give no performances, it remains a living art form. Outsiders would discover some of this outpouring often approaching professional quality, but the local people find these events very enjoyable, even when they are somewhat amateurish, because they know the participants...and that makes all the difference.

Steve comes bounding through the door with Mary at his heels. She had made the mistake of commenting that there were punch and cookies at the show, and Steve of course is completely fulfilled only when he is guzzling punch and chomping cookies. He spots the punch bowl immediately and races across the room to commence his happy ritual.

Jane and Mary decide that since they really will not have time to go home for dinner, they will all stay and have a sandwich at the athletic club. Afterward Mary will make the final sacrifice and take Steve home to put him to bed, even though she had looked forward to seeing Mark in the musical. Following their snack, Mark leaves immediately for the theater while Don, Jane, and Jill take a leisurely walk through the square greeting friends and window shopping.

All the space for the gift shop and other small shops in the square is owned by the homeowners' association. The shop owners pay a percentage of their gross sales as rent. Consequently, these shops are well patronized by the local people. Many of the shop owners are retired but enjoy having something worthwhile to work at, and several of the shops have been considerably more successful and therefore somewhat more time-consuming than the shop owners had originally meant them to be.

The village square is the crossroads of the community where the young people meet their friends and where virtually everyone gets to know everyone else. The square gives the neighborhood a real sense of place and identity. It is the vital nerve center, without which there would be no neighborhood at all.

The theater seats only about eighty people, making the productions intimate, to say the least. Tonight is opening night, so there will probably be a full house. The play will run for two more weekends, and then a new play will be presented. The play tonight is a typical musical comedy, but the actors often ad lib a little, and well-known neighborhood personalities can be assured of getting a roasting. This of course gives the productions a strong local flavor.

Some local movies have also been made and shown here. The first few were pretty dreary, but they have recently improved immensely and are well attended by the young people in particular.

After the play, Don and Jane are congratulated for Mark's performance. (Actually he forgot several lines, but he was quick enough to recover.) More importantly, he pulled off a couple of ad libs that brought down the house. Don and Jane decide to wait for Mark so they can all walk home together. As they move through the warm summer evening they rehash the highlights of the play. Deep inside himself Mark is quite happy, feeling the great satisfaction that all performers experience when they succeed with their audience.

Mary is glad to see them, and after she reports the details of getting Steve to bed, they all gather in the living room pit for a drink. The pit, the Japanese bathtub with space for four people, and the sauna seem to be the natural gathering places for the family. None of them was an expensive addition, and Don often comments that he is certainly glad these elements were part of their house when they bought it.

187

It has been a busy Saturday, but not really too different from many others. Don thinks back to their family's life-style before they moved to the village. He thinks of how Jane has changed. She now feels that she is succeeding as a mother and as an individual, both of which were of deep concern before they moved here. He thinks of Mark and Jill and of how much richer their life is now, and of Steve, who had the good fortune of starting life in the village. As he switches off the lights he notices the television set just sitting there gathering dust. Yes indeed, things really have changed. As he turns the corner to the bedroom, he complains a little to himself about his dining room set and how he is going to have to re-do the whole damn design.

CHAPTER ELEVEN

Conclusion

"Great numbers of inhabitants feel unconnected to either people or places and throughout much of the nation there is a breakdown of community living. In fact there is a shattering of small group life. A number of forces are promoting social fragmentation. We are confronted with a society that is coming apart at the seams."

Vance Packard

One day thousands of years ago, the birds heralded the break of a new dawn and the life of the forest began to stir. Man's predecessor, whoever he was, arose and left his primitive shelter and gathered the necessary equipment for the hunt. His mate, awakened by the cries of their newborn child, gathered the suckling into her arms to provide nourishment. As the sun rose, some of the men gathered in a small hunting band and left the village, while the older men and the younger, inexperienced males remained behind to guard the females and the young. During the morning the women gathered fruits and berries and worked at their primitive looms — but always they watched their children actively engaged in games which were not only a great deal of fun but which contained crucial lessons for survival.

By noon the hunting band had captured their prey and were preparing for the long homeward trek, while back in the village the midday heat had brought virtually all activity to a halt. As the sun began its descent, the hunting band arrived with their bounty, and there was a great rejoicing. After a festive meal where even the very old were fed, the women gathered in their young and returned to their shelter. The men remained for a while rekindling the fire and determining who was to take the night watch, and then as darkness fell, they too returned to the shelter.

Our predecessor found his mate resting on animal skins as the children slept nearby. As he lay down beside her he bragged of his role in the hunt, and she warned him of the animals that were beginning to prowl dangerously close to the village. A soft rain began to fall, creating a peaceful atmosphere as they fell asleep.

This was the pattern of life thousands of years ago. It was the pattern of life yesterday, it remains essentially so today, and it will continue to be the pattern tomorrow. As the sun rises and sets, as the seasons change, the same patterns of behavior are reenacted time after time. When our basic living patterns begin to violate this behavioral chemistry, new and terrifying problems begin to emerge. Those who are gifted in the species somehow survive, but few have the capacity for the rapid adaptability that is now becoming necessary. But the quality of life, more importantly the deep satisfactions gained from everyday occurrences are not evident in the world we are busily creating.

Man evolved slowly. He can change, but the pace is just as slow. If he sees a better way for the future, let him reach out and experiment cautiously. The restructuring of society at today's incredible pace, without a proved basis of direction, can only produce a world reeling out of control. In the gross and ill-conceived experiments we are experiencing today, we are not only violating our past but actually tampering with our survival — and the joy of living is far too often only a memory. Reverence for life is often touted from the pulpit as a great self-effacing virtue, but what is being questioned here is whether or not modern man can achieve a reverence for life for himself and his progeny.

We must control our obsession to conquer nature and learn to accept the natural rhythm of life that has made us one with our natural surroundings. We must accept ourselves as what we are and fully accept our limitations. We are not totally free to choose. As much as we would aspire to rise, we remain so deeply immersed in it that the past almost defies our withdrawal. We are bound by physical laws, but we are also limited by the patterns of living that have served us well in the past. We remain, then, profoundly in need of a territory on which to live and raise our children. Together the pair bond must provide security and education for the young. We also desperately need that community which can provide the family with the necessary security, identity, and stimulation.

Without the help of the neighborhood community, the family is overloaded with excessive demands that families alone do not succeed in fulfilling. The wayward child, the child with deep emotional problems, the child forced out by other family members, and of course the most common and perhaps worst condition of all, the dissolution of the pair bond, are all prevalent examples of the structural collapse of the family. Multi-family, high-rise housing in well-planned communities and single-family housing in unplanned subdivisons remain the half-right answers that are still all wrong. Housing facing on a golf course or artificial lake, or any other kind of single-function open space, represents the ill-conceived pseudo-solutions thrust upon the unsuspecting public. A house with a yard facing a park that leads to the village center remains the simplest and best answer.

We have not been concerned here with the larger problem of the city, or even the larger community outside the neighborhood. These are extremely complex entities with problems for which there are no simple answers. The neighborhood, on the other hand, presents a reasonably simple problem with some very specific solutions. If, however, we succeed in creating well-conceived housing in healthy neighborhoods, many of the problems of the larger community will be solved. Hopefully, one day five thousand years in the future the sun will rise on family territories in healthy, well-organized communities dedicated to the cooperative improvement of their inhabitants. As this generation has inherited from generations past the knowledge of what comprises a healthy human environment, we must diligently strive to reintegrate these timeless elements into a radically changing world and become the critical link in an even brighter future for the species *Homo sapiens*.

Primitive religion concerns itself with the gods, but its teachings revolve around pragmatic concepts that solve the critical problems of survival. Organized religion, as it later evolved, begins to concern itself primarily with the survival of its institutional self rather than with the individual, and thereby becomes immersed in its own irrelevancy.

The new community faith must respond to the crucial problems of survival in a very direct and simple way and thereby fulfill the highest function of religion. A new respect for nature and for the magnificent planet we inhabit, a new respect for this life now, and a new respect for our fellow beings, indeed for all living things that share this life with us, must become the new theology. But in reality this new "religion" is the same old time-worn teaching that has stood us well in the past — and only it can now assure us a future.

Once we recognize man's limitations and fully accept him as he is, then can we begin to discover his incredible potential.

Appendix

The costs involved in housing are continually changing, just as are the costs of all other goods and services. Not only do costs change from time to time, but they also change from place to place. Therefore, it is impossible to assign accurate cost estimates for the concepts which have been presented in the previous pages. However, based on current costs and based on the fact that costs tend to change in some sort of relative way, it is possible to offer some suggestions about the comparative financial aspects of building homes plus an environment.

Most of what has been said in these pages is based on the belief that to change our housing patterns from being marginal, as they are presently, to being forward looking and healthy would require only a 10 percent change in the financial structure of the housing industry. Of that 10 percent, 3 percent would be used for built-in furniture to make the house livable, 3 percent would be used for landscaping to make the yard usable, 3 percent would be used for open space for a place to play, and 1 percent would be used for a village center to give the neighborhood identity. Thus, instead of the cost of a house including only land, house, finance, and profit, it would also include these four additional items.

It is this 10 percent that would make all the difference in the way we live. But at present, the Federal Housing Administration neither requires nor guarantees to insure these items, banks will not finance such items as part of the cost of housing, and city planners will not require them.

The housing industry is dedicated to producing a substandard, unfinished product which is totally vulnerable to the forces which continue to destroy entire housing areas. Those forces are not only physical but social and psychological as well.

Ten percent is all that stands in the way.

This appendix summarizes in tabular form some of the concepts and amenities which could be provided by that 10 percent adjustment.

Requirements of Private Territory (House and Yard)

Individual and Family Needs	The Right Kind of Space	For the Right Things To Happen
Secure spaces and warm rest	A fireplace and conversation pit; a place to dine and not feed; a place for family ritual; an outside fire pit; a summer gazebo	A secure person
Exciting spaces	An atrium; a two-story space; a crow's nest; outside desks, pools, and vistas	An expanded view of life
Learning and creativity	A studio; a music area, a library; a workbench; a creative play area outside; a greenhouse; a vegetable garden	A will to investigate; a will to learn; a will to experiment; the dignity of high-level aspiration
Esthetic	A galley wall; a stereo area; a library; outside sculpture, moon lanterns, and natural landscaping	A rich cultural heritage; a love of beauty

Individual and Family Needs	The Right Kind of Space	For the Right Things To Happen
Spiritual	A place to be alone to contemplate; a shrine to nature	An inner direction; self-identity
Athletic	A vital play space; a place to exercise; a place to fight it out; a place to be free and uninhibited; basketball hoop; swing set, swimming pool	Stimulation; a commitment to good physical condition and a desire to outdo oneself
Rest and other physical needs	A place to relax	Health and well-being

Requirements for Cooperative Territory or Social Space (Park and Square)

Individual and Family Needs	The Right Kind of Space	For the Right Things To Happen
A secure space	A gazebo; a fire pit; benches and playground; picnic areas; the village square	A friendly neighborhood feeling of security; identity in a larger group; community bond
Exciting spaces	The village square; arching trees; open meadow; great vistas; mounding; a duck pond	An expanded view of life
Learning spaces	A great variety of play; community library areas; visiting lecturers; a greenhouse; a hobby house; creative play area; a co-op nursery	A lush learning environment
Esthetic	A community art show; theater; dance; fountain and sculpture; coordinated village symbols	A rich cultural environment
Spiritual	Community government; a place to be alone and watch a sunset; a beautiful setting for living	Tolerance for a variety of views; a deep and abiding love of nature; appreciation of community ritual
Athletic	A sleigh hill; swimming pools; tennis courts; bicycle paths; play in fields; deck chess	Health and well-being
Rest and other physical needs	Shaded benches; a water fountain; public restrooms; an outside cafe; a large outside fire pit	A relaxed, friendly atmosphere, with a deep concern for people

The Nature of the Amenities

The structure of the community is based on the functional demands of the young. Since the maturation process requires ever-expanding parameters, the community structure is a direct response to those requirements.

Structure	Amenities and Functions	
House	Private Yard	
Micro Neighborhood (10–12 families) 3- to 6-year-old children	Open Space Tot Playground Co-op Nursery in Homes	
Neighborhood (100–125 families) 7- to 12-year-old children	Open Space Small Clubhouse Cub Scouts Brownies	Mothers' Clubs Neighborhood Government Local Choice Amenities
Community (500–600 families) 13- to 18-year-old children	Village Center Teen Center Educational Amenities Playing Fields	Tennis Courts Youth Organization Theater Groups Special Interest Groups

Cost of Community

With proper planning it is possible to almost double the number of housing units on a hundred-acre tract and still provide more open space and a healthier living environment, not to mention a community center.

Subdivision of 100 Acres		Planned Neighborhood of 100 Acres	
3.5 Units per Acre		6.5 Units per Acre	
No Park		22-acre Open Space and Playground	
No Community Center		Community Center	
		Athletic Club	
		Shops	
		Hobby Shop	
		Co-op Nursery	
		Town Hall	
350 Total Units		650 Total Units	
House	$20,000	Housing	$20,000
Lot (10,000 square feet)	6,000	Lot (4,320 square feet)	4,000
		Community Cost	2,000
Merchandising	5,000	Merchandising	5,000
Sales price	$31,000	Sales price	$31,000

Community Facilities for 500 Families

	Square Feet	Cost
Community Center ($246,000)		
Meeting Room (A flexible space for town meetings, theater, recitals, etc., including a kitchen and storage)	2,560	$ 35,000
Reception and Party Room (Includes fireplace and conversation pit, plus good lighting for art shows)	1,500	25,000
Games Area and Teen Center (Includes billiards and ping pong)	1,500	25,000
Art Center (Contains art studio, ceramic kilns, etc.)	1,000	12,000
Wood Shop	1,000	10,000
Office, Toilets, Utilities, etc.	900	16,000
Athletic Club		
Indoor Pool (20 x 40 feet)		45,000
Outdoor Pool (30 x 80 feet)		20,000
Dressing Rooms with Sauna	1,200	20,000
Exercise Room with Equipment	600	8,000
Four Tennis Courts		20,000
Open Space ($220,000)		
20 Acres at $5,000 per Acre		100,000
Landscaping and Sprinkler System		80,000
Four Playgrounds		40,000
TOTAL COMMUNITY COSTS		$466,000

The total amounts to approximately $1,000 per living unit. Other amenities which might be considered include an experimental laboratory ($10,000), an observatory ($5,000), a greenhouse ($15,000), and a mechanics center ($8,000).

The costs of a cooperative farm or some form of cooperative industry are difficult to determine. However, those additions might help to pay for other amenities. A condominium complex might also be considered as a part of the community. Although not an amenity, such a complex would help to minimize management costs, because the condominium management can also manage the home owners' association.

The operating costs of the amenities itemized above should run about $70,000 per year or $140 per year per living unit. This amounts to approximately $12 per month.

References

Ardrey, Robert. *Social Contract.* New York: Atheneum, 1970.   *203*

———————. *The Territorial Imperative.* New York: Dell Publishing Co., 1966.

Beauvoir, Simone. *The Coming of Age.* New York: G. P. Putnam's Sons, 1972.

Benedict, Ruth. *Patterns of Culture.* Boston: Houghton Mifflin Co., 1934.

Biddle, William; and Biddle, Loureide J. *The Community Development Process.* New York: Holt, Rinehart and Winston, Inc., 1965.

Breckenfeld, Gurney. *Columbia and the New Cities.* New York: Van Rees Press, 1971.

Bronfenbrenner, Urie. *Two Worlds of Childhood.* New York: Russell Sage Foundation, 1970.

Calhoun, John B. "The Role of Space in Animal Behavior." *Environmental Psychology: Man and His Physical Setting.* H. Proshansky *et al.*, Editors. New York: Holt, Rinehart and Winston, Inc., 1970.

Darwin, Charles. *The Origin of Species.* Chicago: Great Books — William Benton, 1952.

Dubos, René. *A God Within.* New York: Charles Scribner's Sons, 1972.

——————. *Man, Medicine, and Environment.* New York: Mentor, 1968.

——————. *So Human an Animal.* New York: Charles Scribner's Sons, 1968.

Fitch, James Marston. *American Building — The Historical Forces That Shaped It.* Boston: Houghton Mifflin Co., 1966 and 1972 (1947 orig.).

Fuller, R. Buchminster. *Utopia or Oblivion.* New York: Overlook Press, 1969.

Galbraith, John Kenneth. *The Affluent Society.* Boston: Houghton Mifflin Co., 1969.

——————. *The New Industrial State.* Boston: Houghton Mifflin Co., 1967.

Gans, Herbert. *The Levittowners.* New York: Pantheon Books, 1967.

——————. *People and Plans.* New York: Basic Books, Inc., 1968.

Glasser, William. *The Identity Society.* New York: Harper & Row, 1972.

Golding, William. *Lord of the Flies.* New York: Capricorn Books, 1959 (1954 orig.).

Goodman, Paul; and Goodman, Percival. *Communitas.* New York: Random House, 1960 (1947 orig.).

Gordon, Michael. *The Nuclear Family in Crisis: The Search for an Alternative.* New York: Harper & Row, 1972.

Gruen, Victor. *The Heart of Our Cities.* New York: Simon & Schuster, 1964.

Gutman, Robert (Editor). *People and Buildings.* New York: Basic Books, 1972.

Hall, Edward T. *The Hidden Dimension.* New York: Doubleday, 1966.

Hardin, Garret. *Exploring New Ethics for Survival.* New York: Viking Press, 1972.

Hogbin, Ian. "A New Guinea Childhood." *From Child to Adult.* New York: Natural History Press, 1970.

Holt, John. *Freedom and Beyond.* New York: E. P. Dutton & Co., Inc., 1972.

Huxley, Aldous. *Brave New World Revisited.* New York: Bantam Books, 1960 (1958 orig.).

Jacobs, Jane. *The Death and Life of Great American Cities.* New York: Random House, 1961.

Kotler, Milton. *Neighborhood Government.* New York: Bobbs-Merrill Co., 1969.

Kuhns, William. *The Post Industrial Prophets.* New York: Harper & Row, 1971.

Lady Allen of Hurtwood. *Planning for Play.* Cambridge, Massachusetts: M.I.T. Press, 1968.

Landau, Elliott D.; Epstein, Sherrie Landau; and Stone, Ann Plaat (Editors). *Child Development through Literature.* New Jersey: Prentice-Hall, Inc., 1972.

Leyhausen, Paul. "The Communal Organization of Solitary Mammals." *Environmental Psychology — Man and His Physical Setting.* H. Proshansky *et. al.*, Editors. New York: Holt, Rinehart and Winston, Inc., 1970.

Long, Norton, *The Unwalled City.* New York: Basic Books, Inc., 1972.

Lorenz, Konrad. *Kong Solomon's Ring.* New York: Time-Life, Inc., 1952.

Masotti, Louis H.; and Hadden, Jeffrey K. (Editors). *The Urbanization of the Suburbs.* Beverley Hills: Sage Publications, 1973.

McHarg, Ian. *Design with Nature.* New York: Doubleday, 1971.

Mead, Margaret. "How Will We Raise Our Children in the Year 2000?" *Saturday Review* (March 1973).

Melville, Keith. *Communes in the Counter Culture.* New York: William Morrow & Co., 1972.

Morris, Desmond. *The Human Zoo.* New York: Dell Publishing Co., Inc., 1971 (1969 orig.).

_____. *The Naked Ape.* New York: Dell Publishing Co., Inc., 1971 (1967 orig.).

Mumford, Lewis. *The Story of Utopias.* New York: Viking Press, 1972 (1950 orig.).

Newman, Oscar. *Defensible Space.* New York: Macmillan Co., 1972.

Newton, Norman T. *Design on the Land.* Cambridge, Massachusetts: Harvard University Press, 1971.

Packard, Vance. *A Nation of Strangers.* New York: Pantheon Books, 1972.

Papanek, Victor. *Design for the Real World.* New York: Pantheon Books, 1971.

Perin, Constance. *With Man in Mind.* Cambridge: M.I.T. Press, 1970.

Proshansky, Harold M.; Ittelson, William H.; and Revlin, Leanie G. *Environmental Psychology: Man and His Physical Setting.* H. Proshansky *et. al.*, Editors. New York: Holt, Rinehart and Winston, Inc., 1970.

Reich, C. *The Greening of America.* New York: Bantam Books, 1971 (1970 orig.).

Riesman, David. *The Lonely Crowd.* New Haven: Yale University Press, 1961 (1950 orig.).

Schorr, Alvis L. "The Role of Space in Animal Behavior." *Environmental Psychology: Man and His Physical Setting.* H. Proshansky *et. al.*, Editors. New York: Holt, Rinehart and Winston, Inc., 1970.

Skinner, B. F. *Beyond Freedom and Dignity.* New York: Alfred A. Knopf, 1971.

_____. *Walden Two.* New York: Macmillan Company, 1971 (1948 orig.).

Spiro, Melford E. *Kibbutz*. New York: Schocken Books, 1963.

Suttles, Gerald D. *The Social Construction of Communities*. Chicago: University of Chicago Press, 1972.

Swift, Lindsay. *Brook Farm*. New Jersey: Citadel Press, 1973 (1961 orig.).

Time-Life (Editors). *The First Man* (1973); *Life Before Man* (1972). *The Missing Link* (1972); and *The Neanderthals* (1973). New York: Time-Life Books.

Toffler, Alvin. *Future Shock*. New York: Bantam Books, 1971 (1970 orig.).

Turnbull, Colin. *The Mountain People*. New York: Simon and Schuster, 1972.

Von Eckardt, W. *A Place to Live*. New York: Delacorte Press, 1967.

Ward, Barbara; and Dubos, René. *Only One Earth*. New York: W. W. Norton & Co., 1972.

Whyte, William H. *The Last Landscape*. Garden City, New Jersey: Doubleday, 1968.

_____. *The Organization Man*. New York: Doubleday, 1957.

Index

adult communities, 167–69
aged, housing for, 31, 140
amity (A), 12, 135
architects, 118
Arizona, 27
artificial materials, 97–99

Baltimore, 32
bar, kitchen, 42; wet, 42, 44
Bay Area, 60–61, 79
Beacon Hill, 27
behavioral modification, 17–19;
    positive patterns, 44; sink, 15
biological patterns, 10–14, 40
Boston, 27
builders, 116
built-ins, 64, 92–93, 94

California. See Southern California.
    Also see Bay Area and San
    Francisco
Cape Cod cottages, 27, 67, 68
carpets, 98
Children's Art Gallery, 33
children's rooms, 45

Colonial Spanish, 61
Colorado, 27
Columbia, Maryland, 32, 33, 131–32
commune, 136
community
    all adult, 167–69
    bond, 134–37
    center, 140
    educational, 161–63, 164–67,
        169–73
    healthy, 46–52, 135, 140
    recreational, 161–63
condominium, 140
conversation pit, 42
cooperative territory as "social
    space," 12–14
coordination, 119–23
costs in housing, 197–202

decorating, interior, 74–79, 80–81;
    low-cost housing, 80–81
definition of total environment, 31
Denmark, 27
design, 67–68; Scandinavian, 78
dining room, 42, 65
District of Columbia, 27, 32

Early American, 60, 77
educational communities, 161–63,
   164–67, 169–73
enmity (*E*), 12, 135
environment, total, definition, 31
equation (*A = E + H*), 12–13, 135
España Mediterranean, 60
European
   furniture, 77; designers, 78
   public living, 84–85
   towns, 47, 130
experimental creative area, 82–83

family room, 42–44
Federal Housing Authority, 33–34,
   63–64, 110, 113–16, 121, 122
financial institutions, 111–12
Finland, 29
fireplace, 41–42, 65, 98
"fit," 16 17, 61, 136
formal and informal, 65–67
French Colonial mansions, 27;
   Provincial furniture, 60
function, 65–67
furniture designers, 78; types, 60,
   77–78
future shock, 17, 137

Georgetown, 27
Georgian homes, 27
Great Britain, 29

Harlem, 131
hazard (*H*), 12, 135
high-rise housing, 64
hobby workshop, 141
Homo sapiens, 10, 19, 78, 85, 131, 155
housing
   costs, 197–202
   European, 29, 47, 77, 78, 84–85,
    130
   high-rise, 64
   technology, 100–03
   types (regional), 27–31, 59,
    60–62
human storage, 25, 28

Idaho, 131
Iks of Uganda, 160–61
in-group, out-group, 137, 142

interior decoration, 74–79, 80–81
Italian Provincial, 60, 77

kibbutz, 136
kitchen bar, 42

labor unions, 117
land developers, 29–30, 110–11
landscaping, 81–83
living room, 41–42
local government, 118–19
London, 63; County Council, 120
Los Angeles, 33, 131
low-cost housing, decorating in,
   80–81

Maryland, 32, 33, 131–32
Maybeck house, 75
Mediterranean furniture, 77; style of
   decorating, 79–80
Mexican folk art, 79
mobile homes, 12, 31, 64, 101, 118
Monticello, 73

National Home Builders
   Association, 117
natural setting, 95–96
neighborhood
   healthy, 46–52, 135, 140
   homeowners' association, 143,
    179–90
New England, 27, 79; Guinea,
   157–59; Jersey, 29; Mexico, 27;
   Orleans, 27

open space, 48, 83–85
Operation Breakthrough, 115;
   Turnkey, 113
orbit, 14–16
Oriental Modern, 60

pair bond
   chimpanzee, 153–54
   educating offspring, 40
   evolution of, 10–14
parlor, 77
plumbing, 93
Pocatello, Idaho, 131
Prairie House, 29, 61
prefabrication, 63, 100–02

primitive societies, 157–61
private territory, 14, 40, 41, 51
Puerto Rico, 59–60

Radburn, New Jersey, 29
recreational communities, 161–63
regional design, 27, 60–61, 79–80;
    housing, 27–31, 59, 60–62
Reston, Virginia, 32, 33

"salt box" house, 27
San Francisco, 27, 60, 75
scale, 14–16, 137–43
Scandinavian designers, 78
Sea Ranch, 60
Southern California
    conversation pit, 42
    folk art, 79
    market, 64
    Spanish-style houses, 27
    types of architecture, 60–61
    unfinished houses, 75
    Westlake, 33
    wet bar, 42, 44
Spanish-style houses, 27
subdivisions. *See also* land developers
    initial stages, 29–30, 31
    no open spaces, 48, 84

Sweden, 29

technology, housing, 100–03
territory
    chimpanzee, 153–54
    cooperative, 12–14, 46–52
    man's need for, 10–14
    private, 14, 40, 41, 51
tile, 98
touch-plate electrical wiring, 93, 94
tree, cost of moving, 94

Uganda, 160–61
un-neighborhood, 13, 132, 141, 143
Utah, 27

vacuum system, built in, 92–93, 94
Victorian bric-a-brac, 27
Virginia, Reston, 32, 33;
    Williamsburg, 27, 62

Washington, D.C., 27, 32
Westlake, Southern California, 33
wet bar, 42, 44
Williamsburgh, Virginia, 27, 62
Wogeo tribe of New Guinea, 157–59

209

HOUSE, PLUS ENVIRONMENT was composed in Baskerville, Baskerville Italic, and Helvetica by Twin Typographers. Design was by Annegret Ballingham, and mechanicals were prepared by Fran Clements of Bailey-Montague & Associates. The book was printed by Paragon Press on Crown Matte from Zellerbach Paper Company. Binding was by Mountain States Bindery.